Sensors

Topical Conference at the 2009 AIChE Annual Meeting

Nashville, Tennessee, USA
8-13 November 2009

ISBN: 978-1-61567-918-8

Printed from e-media with permission by:

Curran Associates, Inc.
57 Morehouse Lane
Red Hook, NY 12571

Some format issues inherent in the e-media version may also appear in this print version.

Printed by Curran Associates, Inc. (2010)

For permission requests, please contact AIChE
at the address below.

AIChE
3 Park Avenue
New York, NY 10016-5991

Phone: (203) 702-7660
Fax: (203) 775-5177

www.aiche.org

Additional copies of this publication are available from:

Curran Associates, Inc.
57 Morehouse Lane
Red Hook, NY 12571 USA
Phone: 845-758-0400
Fax: 845-758-2634
Email: curran@proceedings.com
Web: www.proceedings.com

TABLE OF CONTENTS

Proton-Conducting Oxide Thin Film Coated Fiber Optic Sensor for High Temperature Hydrogen Monitoring

Xiling Tang, Department of Chemical & Materials Engineering, University of Cincinnati, Cincinnati, OH

Kurtis Remmel, Department of Chemical & Materials Engineering, University of Cincinnati, Cincinnati, OH

Junhang Dong, Department of Chemical & Materials Engineering, University of Cincinnati, Cincinnati, OH

We report the integration of nanocrystalline perovskite thin film with a structured optical fiber for *in situ* monitoring of bulk hydrogen in conditions relevant to the H_2 production by biomass and fossil fuel gasification and water gas shift (WGS) processes. Hydrogen production from renewable bio-renewable and fossil stocks is in the center of the development of clean power systems. The key constituents in the syngas by gasification include H_2, CO, CO_2, H_2O, light hydrocarbons (i.e. CH_4 etc.), and common impurity H_2S. The ability of *in situ* gas monitoring is imperative to the realization of intelligent process control for improving the energy efficiency, system reliability, and pollutant emission reduction. However, currently available semiconductor and electrochemical sensors are difficult to deploy and operate directly in the harsh environment involving high temperature, corrosive and dusty gases.

In this study, the proton electron/hole mixed conducting perovskite-type $Sr(Ce_{0.8}Zr_{0.1})Y_{0.1}O_{2.95}$ (SCZY) nanocrystalline thin film was synthesized by a facile polymeric precursor route on the long-period fiber grating (LPFG). At high temperature, the defect states in the SCZY crystal lattice change upon reacting with hydrogen that varies its refractive index and light absorbing characteristics. The transmission spectrum of the LPFG is a sensitive function of refractive index of the overcoat (n^{ov}). The SCZY-coated LPFG can thus detect hydrogen by straightforward measurement of its resonant wavelength (λ_R) shift. The sensor's H_2-selectivity, sensitivity, stability, and reversibility have been investigated at 500 ^{O}C and atmospheric pressure in binary and multicomponent gas streams.

See more of this Session: Environmental, Catalytic, and Industrial Sensors
See more of this Group/Topical: Topical 9: Sensors

Non-Biological Inhibition Based Sensing (NIBS) Demonstrated for the Detection of Toxic Sulfides and Arsines

Chelsea N. Monty, Chemical and Biomolecular Engineering, University of Illinois at Urbana-Champaign, Urbana, IL

Nicolas J. Londoño, Chemical and Biomolecular Engineering, University of Illinois at Urbana-Champaign, Urbana, IL

Rich I. Masel, Chemical and Biomolecular Engineering, University of Illinois at Urbana-Champaign, Urbana, IL

The purpose of this work is to demonstrate a new type of sensor, where inhibition of a catalytic reaction is used as a way to chemically amplify the analytical signal from an analyte molecule. By way of background, Blaedel pioneered the idea of using chemical amplification to increase the sensitivity of an analytical measurement. In this paper we have developed a new approach to chemical amplification. Instead of using the analyte as a catalyst, we use the analyte as an inhibitor for a given reaction. In detail, we find reactions A"_B that are catalyzed by a catalyst C, where C selectively binds an analyte of interest. A small change in the concentration of the analyte produces a large change in the concentrations of A and B. Consequently, the signal from the analyte is amplified. We call this technique Non-biological Inhibition Based Sensing (NIBS), because we are using the inhibition of a non-biological based catalysts as a sensing mechanism.

There are many possible catalysts that one can use; enzymes, for example, work well for chemical amplification but suffer from shelf life issues. In place of biomolecules, we have been concentrating on using materials found in Draeger tubes and other related colorimetric methods as catalysts. Draeger tubes are already reasonably selective to analytes of interest, but are not particularly sensitive. We hypothesized that if we use the chemicals in a Draeger tube or related colorimetric methods as a catalyst in a chemical amplification system, we could retain the selectivity of the colorimetric method with a much higher sensitivity. In the work here we demonstrate the success of this technique for the detection of chloroethyl methyl sulfide (CEMS) and diphenylaminechlorarsine. We find that we can enhance the sensitivity of these systems by a factor of two with no apparent loss in selectivity. This paper will focus on optimization of reaction kinetics and analysis of the inhibition reactions.

See more of this Session: Environmental, Catalytic, and Industrial Sensors
See more of this Group/Topical: Topical 9: Sensors

Pd/TiO_2 Nanofibrous Membrane and Its Application in Hydrogen Sensing

Wenzhao Jia, Chemical, Materials and Biomolecular Engineering, University of Connecticut, Storrs, CT

Liang Su, Chemical, Materials and Biomolecular Engineering, University of Connecticut, Storrs, CT

Yu Ding, Chemical, Materials and Biomolecular Engineering, University of Connecticut, Storrs, CT

Ashley Schempf, Chemical, Materials and Biomolecular Engineering, University of Connecticut, Storrs, CT

Ying Wang, Chemical, Materials and Biomolecular Engineering, University of Connecticut, Storrs, CT

Yu Lei, Chemical, Materials and Biomolecular Engineering, University of Connecticut, Storrs, CT

The Pd/TiO_2 nanofibrous membrane was successfully fabricated. The as-prepared membranes were characterized by field-emission scanning electron microscopy (FESEM), energy-dispersive X-ray spectrometry (EDX), X-ray diffraction (XRD) and transmission electron microscopy (TEM). The Pd/TiO_2 nanofibrous membranes show good mechanical and conductive properties, and offer an excellent sensing platform. Its application for hydrogen detection is demonstrated and the developed hydrogen sensor exhibits fast response, excellent recovery behavior, and good sensitivity. Two different hydrogen sensing mechanisms were realized in Pd/TiO_2 nanofibrous membranes: resistance increase caused by the formation of palladium hydride and resistance decrease due to lattice-expansion-induced "break-junction".

See more of this Session: Environmental, Catalytic, and Industrial Sensors
See more of this Group/Topical: Topical 9: Sensors

Highly Sensitive Multiphase Micro-Sensor as Micro-Gas Chromatography Detector for Selective Detection of Toxic Phosphonates

Chelsea N. Monty, Chemical and Biomolecular Engineering, University of Illinois at Urbana-Champaign, Urbana, IL

Nicolas J. Londoño, Chemical and Biomolecular Engineering, University of Illinois at Urbana-Champaign, Urbana, IL

Adarsh D. Radadia, Chemical and Biomolecular Engineering, University of Illinois at Urbana-Champaign, Urbana, IL

Robert D. Morgan, Chemical & Biomolecular Engr, University of Illinois at Urbana-Champaign, Urbana, IL

Rich I. Masel, Chemical and Biomolecular Engineering, University of Illinois at Urbana-Champaign, Urbana, IL

The objective of this paper is to report optimization and integration of a multiphase MEMS- based phosphonate sensor that utilizes selective, liquid chemistry in order to detect toxic phosphonate compounds eluting from a microfabricated gas chromatography (ì-GC) column. Previous work from our laboratory demonstrated that phosphonate molecules can be detected with such a device, but the sensitivity was insufficient for certain applications (e.g. detection of pesticides in foodstuffs) and the response time of the detector is too long for the desired separation with a ì-GC column.

In this paper temperature and injector assembly pressure were optimized and the selectivity of the sensor was tested with interferent mixtures. We find that i) the current sensor design is more sensitive and more selective than previously developed detection methods, ii) the peak width of the optimized sensor makes it a viable ì-GC sensor, and iii) the sensor can detect target molecules from complex mixtures that non-selective detection methods cannot distinguish due to peak overlap. The optimized device can selectively detect 109 molecules of acetylcholinesterase inhibitors (most notably toxic phosphonates) with a minimum peak width of 0.2 seconds.

See more of this Session: Environmental, Catalytic, and Industrial Sensors
See more of this Group/Topical: Topical 9: Sensors

Membrane Catalysts for Optical Sensing of Anhydrides in the Gas Phase

Subasri Ayyadurai, Chemical and Materials Engineering, University of Cincinnati, Cincinnati, OH

Jonathan A. Bernstein, Division of Immunology, University of Cincinnati Medical School, Cincinnati, OH

Anastasios Angelopoulos, Chemical and Materials Engineering, University of Cincinnati, Cincinnati, OH

The diffusion of low molecular weight dyes from dilute solutions with alcohol into transparent ionomeric membranes is being investigated to create unique supramolecular structures which respond optically to their environment. The membranes act as both durable dye supports and catalysts to sensitize the dyes to the presence of organic chemical hazards in the gas phase. In this paper, alteration of fluorescence dye concentration within the hydrophilic/hydrophobic domains of ionomeric membranes is shown to control sensitivity and selectivity toward gas phase exposure of various anhydrides which are considered to be occupational hazards in the polymers industry. Trimellitic, phthalic, and maleic anhydride are investigated at temperatures above the melting point of the solids. UV/VIS and FTIR are employed to identify Friedel-Crafts acylation as the sensing mechanism as well as to quantify the diffusion and solubility of the dyes and the anhydrides within the membranes. Appropriate choices of solvent for dye impregnation as well as the carrier gas for the anhydrides are shown to be critical in achieving *parts per billion* sensitivity and selectivity toward simultaneous exposure of the three anhydrides.

See more of this Session: Environmental, Catalytic, and Industrial Sensors
See more of this Group/Topical: Topical 9: Sensors

Detection of Gases On Biosensor Surfaces

Ajit Sadana, Chemical Engineering Department, University of Mississippi, University, MS

Ifejesu Eni-Olorunda, Chemical Engineering, University of Mississippi, University, MS

A fractal analysis is presented for the binding and dissociation (whereever applicable) of different gases on different biosensor surfaces. Both single- and dual-fractal analysis are used. The dual-fracatal analysis is used only when the single-fractal analysis did not provide an adequate fit. Corel Quattro Pro 8.0 is used to fit the data. The fractal analysis is used to analyze (a) the binding of liquid petroleum gas (LPG) to zinc oxide films prepared by the spray pyrolysis method onto a glass substrate (Shinde et al., 2007), (b) the binding and dissociation of different ammonia concentrations in air to a sol-gel derived thin film biosensor (Roy et al., 2005), (c) binding and dissociation of different methanol concentrations (in ppm) to a polyimide thin layer biosensor (Manera et al., 2002), and binding to a nc-Fe3O4/Si-NPA (nanocrystal magnetite silicon nanoporous array)(Wang and Li, 2005).

The binding rate coefficients are quite sensitive to the degree of heterogeneity or the fractal dimension on the biosensor surface, for example, (a) the zinc oxide films sprayed on glass substrates for the detection of 0.2 volume percent LPG in the gas phase, (b) the binding of ammonia to a sol-gel derived thin film biosensor where the influence of pre-sintering temperature was analyzed,(c) the binding of different concentrtaions of ppm methanol in air to a polyimide thin layer biosensor (Manera et al, 2007), and and the influence of different carrier gases (nitrogen, air, and argon) for the detection of ammonia by an ammonia sensor.

The detection of different gases present in air and in different carrier gases using different types of biosensors provides one with an idea of what is available in the literature and what gases may de detected. Novel biosensor techniques are continuously being developed for the detection of even conventional gases such as ammonia. Hopefully, these novel biosensor techniques should help detect conventional and other gases at lower and lower detection levels; besides becoming more and more sensitive, robust, and reliable detection techniques.

See more of this Session: Environmental, Catalytic, and Industrial Sensors
See more of this Group/Topical: Topical 9: Sensors

Invited 1

Robert Savinell, Chemical Engineering, Case Western Reserve University, Cleveland, OH

The honoree of this symposium, Professor Chung Chiun Liu, is well known for his contributions and innovations over a career to the field of microfabricated chemical and biochemical sensors. In more recent years he also has made substantial contributions to microfabricated energy storage and conversion devices for battery-replacement and for powering sensor devices, and I will summarize some of Liu's approaches and contributions in this field, including more detailed discussions on our collaboration developing micro-fabricated hydrogen fuel cells. These devices have the decoupled energy/power advantages of fuel cells, the simplicity of batteries, and the high power pulse capabilities of ultracapacitors.

See more of this Session: Micro and Nanofabricated Sensors. In Honor of CC Liu I
See more of this Group/Topical: Topical 9: Sensors

MICROFABRICATED SENSORS AND NANOMATERIALS FOR IMPROVED HYPERGOLIC CHEMICAL DETECTION IN MISSILE CANISTERS

Dr. Benjamin J. Ward, Makel Engineering, Inc., Cleveland, OH

INTRODUCTION

Hypergolic fuel/oxidizer systems are broadly deployed in rocket and missile systems, including those for NASA and Ballistic Missile Defense Systems (BMDS). Safety considerations and the need to maximize system operational readiness is pushing the continued development of hypergolic fuel and oxidizer sensor systems for use in missile transport/launch canisters. Makel Engineering is developing a smart sensor system that takes advantage of microfabrication techniques and integrates solid state gas detection films on integrated sensor platform structures. A smart sensor which can operate in a sealed missile canister and can discriminate a significant leak from background interferences will provide the end-user with intelligent, actionable information regarding the presence and significance of a leak so that proper corrective action can be taken. A low power, modular system with long sensor element life will result in a low maintenance system design for minimum logistical footprint.

BACKGROUND

Hypergolic Fuels/Oxidizers

Hydrazine (including monomethyl hydrazine and unsymmetrical dimethylhydrazine) and other hypergolic fuels have a long history of use as rocket propellants, and a wide variety of fuel/oxidizer combinations can be used [1], depending on the propulsion requirements and safety considerations of the target application. Hydrazine has been utilized more broadly in chemical processing, fuel cells and as a corrosion inhibitor in industrial boiler water. Hydrazine is a neurotoxin and produces carcinogenic and mutagenic effects. Accidental hydrazine poisoning induces vomiting, severe irritation of the respiratory tract and long-term effects on the central nervous system. Furthermore, it is believed to be oxidized by peroxidases in the body into reactive intermediates capable of causing a variety of adverse side effects including DNA damage.

Common hypergolic oxidizers include Red Fuming Nitric Acid (RFNA), Mixed Oxides of Nitrogen (MON) and nitrogen tetroxide (NTO). NTO exists as an equilibrium of N_2O_4 and NO_2. MON is a solution of Nitric Oxide (NO) in Nitrogen Tetroxide/Nitrogen Dioxide (N_2O_4/NO_2). In addition to being a strong oxidizer which could attack system components in the event of a leak, human exposure to high levels of NO_2 can cause sudden death, pulmonary edema (resulting in asphyxiation), and/or bronchiolitis obliterans if inhaled, and skin or eye irritation if physical contact is made.

Hypergolic Detection Technologies

Numerous detection technologies have shown sensitivity to hypergolic compounds, but are often limited to specific ambient conditions. Aromatic aldehydes [2] and gold salts [3] have been used in hydrazine indicator badges, as they show a color change upon exposure to HZ or MMH. These work well for passive or wearable exposure detectors, but color change indicators are not well suited for monitoring the interior of a sealed missile canister. Draeger tubes filled with an indicating reagent can be used, but would require access to a sealed canister and would require personnel to flow gas into the tube and read a result.

Chemically modified electrodes have been developed in response to the tendency of hydrazines to require an overpotential when performing electrochemical measurements via redox reactions. Such electrodes have shown promise in electrochemical measurement of hydrazine concentration when operated in a medium that allows the absorption of hydrazine. [4] This introduces a difficulty for operating in dry nitrogen, as the typical gel or polymer mediums require some level of hydration to operate.

Wet electrochemical sensors have been implemented, but have limitations on operational lifetime, ambient temperatures at which they can operate, and possible leakage of electrolyte. Optical systems have been employed as well [5]

Solid State Microsensors for Hypergolic Detection in Missile Canister Environments

Solid state, vapor sensitive materials detectors have been investigated [6], but selectivity and complex detection schemes can be difficult to overcome. Palladium and its alloys are candidate sensing materials which indicate the presence of hydrogen and hydrogen containing species (including HZ, MMH, etc.) by catalysis of hydrogen which results in a resistance change. This reaction is highly selective to hydrogen and hydrazines, and can be monitored using relatively simple measurement techniques. Similarly, tin oxide nanomaterials have exhibited high sensitivity to reducing and oxidizing gases and can be tailored for operation in a nitrogen-purged canister environment.

Solid state, integrated microsensors also offer the potential for flexible operation with smart electronics (detection limits, alarm limits).

EXPERIMENTAL

Sensor Design and Fabrication

The ultimate goal of this development is a smart, integrated sensor capable of standalone operation for extended periods. Since the chemical interactions between the target vapor and the sensor surface are temperature dependent, closed loop temperature control is a necessity. The fuel and oxidizer sensor structures, shown in Figure 1 and Figure 4 respectively, include a temperature detector, heater, and sensing element. Microfabrication and thin film processing allow the overall sensor package to be small in size with low heat loss and minimal energy consumption. Energy consumption can be further reduced by mounting the device over an air gap so that the sensor components (temperature detector, heater, and sensing element) are less affected by conduction heat losses. This decreases power consumption for heating and decreases the time for thermal equilibrium should the environment change. The temperature detector and heater are sputter-deposited platinum films and can operate over a wide temperature range.

Hydrazine Sensor Structure

The hydrazine sensor consists of a highly sensitive Schottky diode structure incorporating a palladium alloy metal gate on doped silicon, a platinum RTD, and a platinum heater resistor. The palladium metal serves as the catalyst for molecular hydrogen to dissociate from hydrazine/MMH/UDMH. The atomic hydrogen then diffuses into the palladium alloy causing a change in the electrical characteristics of the Schottky barrier. When exposed to hydrazines, a layer of Pd-H dipoles forms at the metal insulator interface and changes the depletion layer charge distribution. Changes in electrical characteristics of the device reflect the level of ambient hydrazine up to the point that all absorption sites have been filled. A secondary effect of hydrazine reacting with surface oxides may contribute to sensor response, but has not been investigated here.

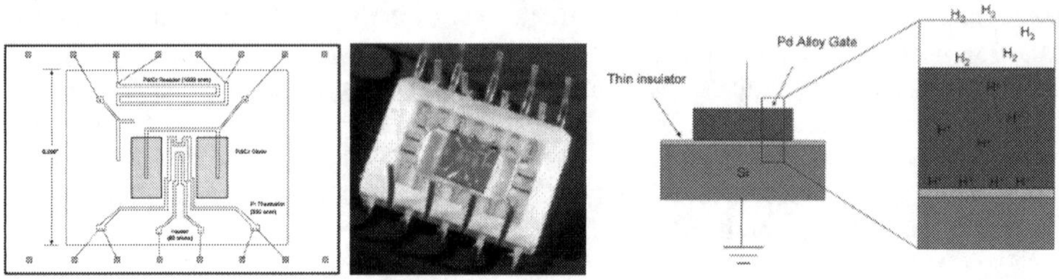

Figure 1. Wide Range Pd-alloy Hydrogen Sensor (left to right, 2-D CAD layout, sensor mounted on ceramic carrier, and Schottky diode detection mechanism to hydrogen-containing species.

N_2O_4/NO_2 Sensor Structure

Makel Engineering and our research partners have demonstrated tin oxide based sensors that are highly sensitive to NO_2. The tin oxide sensing element is composed of interdigitated electrode elements, shown in across which SnO_2 is deposited by spin coating a sol-gel precursor followed by thermal processing. The sensor incorporates a stabilized nanocrystalline SnO_2 layer, using a sol-gel process originallay developed at Case Western Reserve University [7], shown in Figure 2 and Figure 3. The grain size of the SnO_2 grains shown in Figure 3 are approximately 10 nanometers. Nanocrystalline materials have several inherent advantages over conventionally fabricated materials including increased stability and sensitivity at high temperature. Changes in conductivity of doped SnO_2 across the interdigitated electrodes is measured and correlated to gas species concentration.

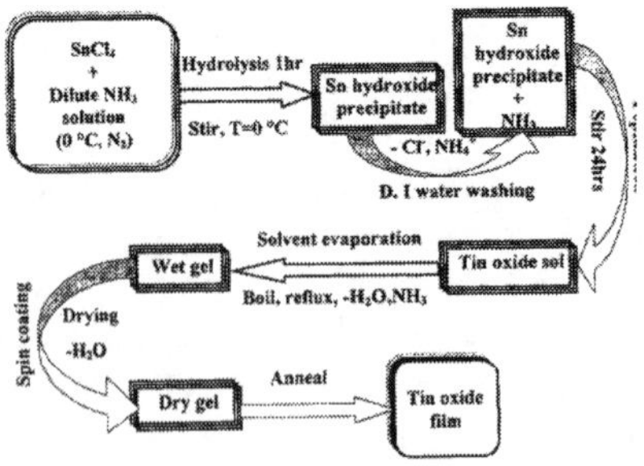

Figure 2. Sol-Gel Process to Produce Stable Nanocrystalline Thin Films.

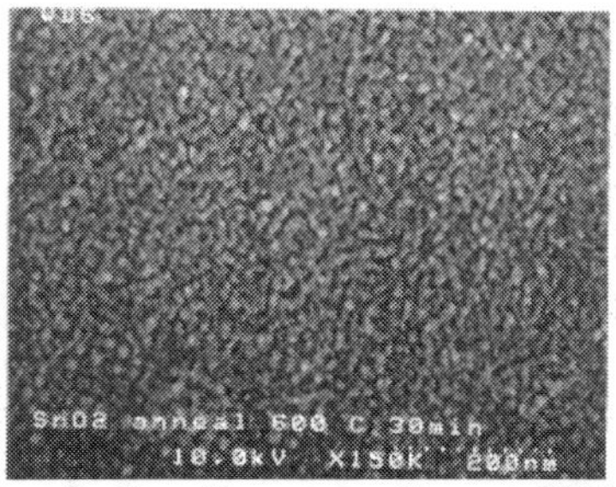

Figure 3. Nanocrystalline SnO_2 after annealing at 600°C for 30 minutes.

Chemical detection with a tin oxide based sensor is based on the effect of chemisorbed species and the resulting effect on the conductivity of the sensing layer. Gas species that undergo oxidation processes increase the conductivity of the tin oxide layer and gas species that are reduced by the layer decrease the conductivity. We have demonstrated gas sensors using this microfabricated and micromachined fused-silica platform structure in a variety of applications [8]. The sensitivity of the sensor for different species can be "tuned" by adding catalytic material to the SnO_2 layer.

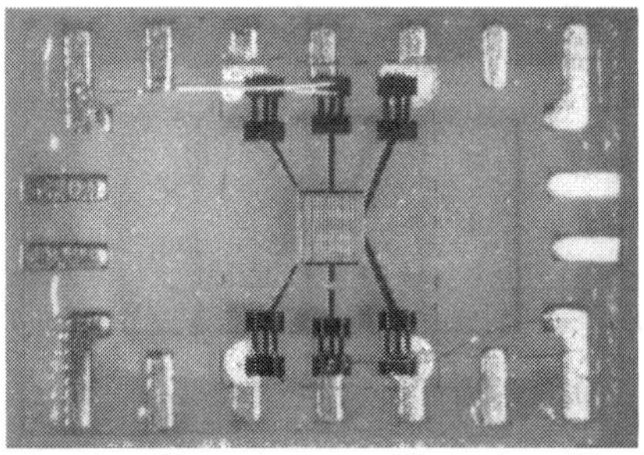

Figure 4. SnO_2 Sensor Including Temperature Detector, Heater and Sensing Element

Laboratory Test Environment

A custom gas exposure test apparatus was developed due to the highly toxic properties of the hypergolic chemicals to be tested and the requirements for highly accurate control of the level of exposure to the sensors. Kin-Tek permeation tubes loaded with MMH or N_2O_4 were used during testing. The exposure concentration was controlled by altering the flow of nitrogen dilution gas through the permeation cell and test manifold. The gas concentration is given by the equation:

$$ppm = K_0 * ng/min / F$$

where ppm is the gas concentration, K_0 is the permeation constant provided by Kin-Tek, and F is the dilution flow rate. Calibration data for permeation rates of each species through the permeation membranes at a specified temperature

A Delrin flow manifold was machined with sensor ports and gas inlet/outlet fittings, and sensor test articles were mounted to a Delrin back-plane for mounting. Any exposed metal surfaces were covered in epoxy if possible, in order to minimize interaction with MMH. The full system was operated within a fume hood with the exception of the nitrogen tank and the PC used for data acquisition.

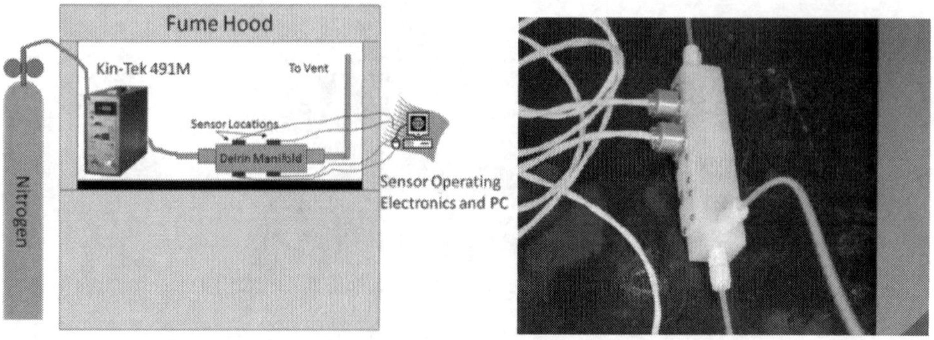

Figure 5: Test setup and flow through sensor manifold for hypergolic chemical testing

RESULTS

Sensor Response Data to Hydrazine

The palladium alloy sensors provided sensitivity and selectivity to MMH to levels as low as 5 ppm (lowest tested concentration). As is typical of chemical sensors, there is a shoulder in the sensor response vs. exposure time curve at low concentrations. The response curve appears much more like a step change as concentrations increase, which is seen Figure 6, Figure 7, and Figure 8. In order to provide the earliest detection capability, it may be preferred to use the initial sensor response slope to predict the MMH concentration level as there is a well defined relationship between the initial slope and the actual MMH concentration.

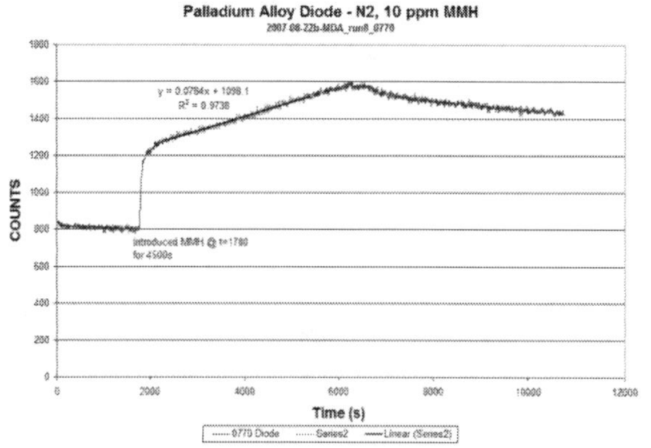

Figure 6: 10 ppm exposure to MMH

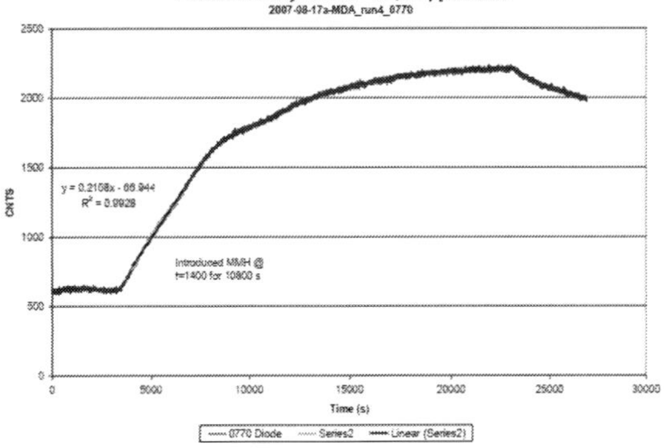

Figure 7: 96 ppm exposure to MMH

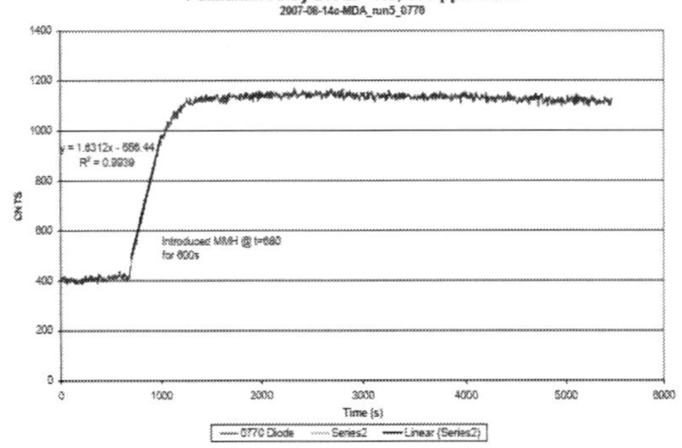

Figure 8: 300 ppm exposure to MMH

Sensor Response Data to N_2O_4 (NO_2)

Nanocrystalline tin oxide sensors provided sensitivity and selectivity to NO_2 to levels as low as 3 ppm (lowest tested concentration). The tin oxide sensors are typically operated at elevated temperature to enhance their reactivity, which is evidenced in improved sensitivity and time response. However, heating the sensors can cause concern that they might act as an ignition source and also results in extra power draw during sensor operation. the tin oxide sensors were operated at 200°C, 100°C, and room temperature (approximately 22°C), shown in . While the fastest time response and sensitivity to NO_2 did come from the sensors operating at 200°C, all three operating temperatures showed suitable sensitivity. Time response can be enhanced by signal processing algorithms that monitor rate of change of the sensor.

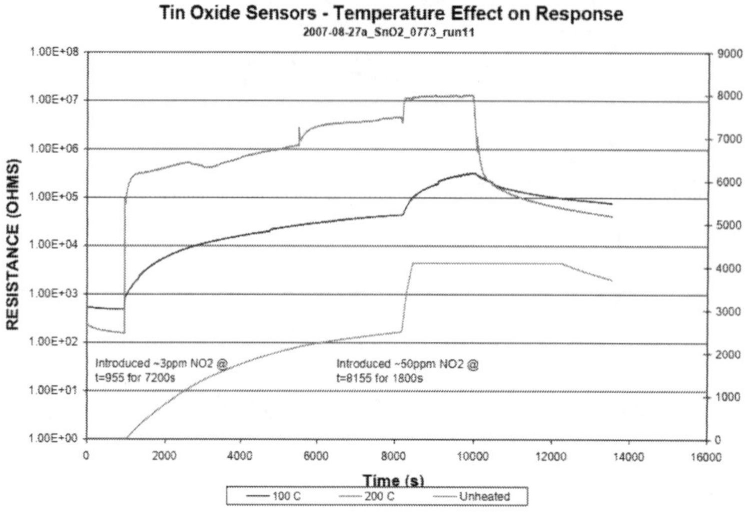

Figure 9: Sensor response of the tin oxide sensors at various operating temperatures

Cross Sensitivity and Background Effects

For the palladium alloy sensor, cross sensitivity to NO_2 is minimal, with a possible small shift in baseline during exposure to NO_2. However, this baseline shift may in fact be due to the sensor temperature adjustments after a change in gas flow rate from the permeation cell. In either case, the magnitude of the shift is very small relative to the sensor signal, and is opposite to an MMH exposure event.

Tin oxide is reduced by MMH, but is also reduced by dry nitrogen background, so the cross sensitivity to MMH does not affect operation in a canister environment. More importantly, the sensor element's NO_2 response is significantly stronger even at low ppm levels than MMH effects

Minimal sensitivity to ammonia (clutter interferent from RTV sealants) was seen at typical operating temperatures.

Reversibility

Though not shown here, the responses of both the palladium sensor and nanocrystalline tin oxide are reversible. Typical of gas sensors, desorption/recovery is slower than the adsorption/catalysis of the gas of interest. The reversibility provides a significant advantage for long term operation. Unlike a colorimetric sensor, the response is not a function of the cumulative exposure but of the ambient concentration of the detected species. This means that even in the presence of a low level background concentration of hypergol (e.g. from assembly of a missile/canister system), the sensor will not false alarm and can still detect potential sudden leaks.

CONCLUSIONS

Based on initial screening data, microfabricated, solid state sensors incorporating palladium alloy sensing elements for hydrazines and nanocrystalline tin oxide sense elements for NO_2 boil-off from N_2O_4 oxidizers meet detection requirements for use in dry-nitrogen purged containers. Solid state chemical sensors operate in missile container environment (absence of oxygen and humidity, presence of clutter interferents from missile canister outgassing) and are better suited than optical systems to withstand shock/vibration of missile canister (no optical misalignment, etc.). Additional development to improve sensor packaging and operation, with a target of an integrated smart sensor system for BMDS applications, is ongoing.

ACKNOWLEDGEMENTS

The author would like to thank fellow researchers at Makel Engineering, Inc. and the Electronics Design Center at Case Western Reserve University. The author's work described here is supported by the United States Army Space and Missile Defense Command under SBIR Grant W9113M-08-C-0174.

REFERENCES

1. Thunnissen, D., Guernsey, C.S., Baker, R.S., and Miyake, R.N. "Advanced Space Storable Propellants for Outer Planet Exploration" 40th AIAA/ASME/SAE/ASEE Joint Propulsion Conference and Exhibit 11-14 July 2004, Fort Lauderdale, Florida AIAA 2004-3488
2. Taffe, P.A. and Rose-Pehrsson, S. L. United States Patent, 4,900,681
3. "Precious-Metal Salt Coatings for Detecting Hydrazines" NASA Tech Briefs, December 2004
4. Sun, C., et al. "Fabrication of a multilayer film containing cobalt phthalocyanine on the surface of a gold electrode based on electrostatic interaction and its application as an amperometric sensor of hydrazine" Thin Solid Films 288 (1996) pp. 291-295
5. Houseman, J. Webster, C. Jackson, J. Brown, P. "Hydrazine Detection with a Tunable Diode Laser Spectrometer" JANNAF PDCS/S&EPS Joint Meeting San Diego, California, USA 26-Apr-1999
6. Holmstrom, S. D. et al. "Mediated Oxidation and Determination of Gaseous Monomethyl Hydrazine in a Solid-State Voltammetric Cell Employing a Sol-Gel Electrolyte" Electroanalysis 12, No. 4 (2000) pp.262-266
7. Jin, J., Zhou, H. J., Jin, Z. L., Savinell, R. F., and Liu, C. C. "Application of Nano-crystalline Porous Tin Oxide Thin Film for CO Sensing." Sensors and Actuators B, 52 (1998) pp. 188-194.
8. Hunter, G.W., Liu, C. C., and D. B. Makel, "Microfabricated Gas Sensors for Aerospace Applications," The MEMS Handbook, CRC Press, Ed. Gad-el-Hak, M, pp.22-1-24, 2002

Determining the Levels of Key Liver Biomarkers Using An Ir/C Printed Biosensor for the Detection of Liver Disease

Brandon Bartling – Department of Chemical Engineering, Case Western Reserve University
Chung-Chiun Liu – Department of Chemical Engineering, Case Western Reserve University
Lu Li – Department of Chemical Engineering, Case Western Reserve University

Abstract - A growing problem in today's population is the increasing number of individuals suffering from forms of liver disease. Detection and diagnosis of diseases, such as hepatitis and cirrhosis, would benefit from the development of a rapid and easy detection method. A method would be especially vital in economically disadvantaged countries such as Egypt, Rwanda and Tanzania, where more than 10% of the population is suffering from some form of liver disease.

When an individual becomes infected with liver disease changes in a series of chemicals in the body, called biomarkers, are observed. Two particular biomarkers that have shown good clinical correlation to the presence of liver disease are total bile acid and adenosine deaminase activity. By monitoring the levels of these biomarkers in the body it is possible to detection the presence of liver disease, and to some level the type.

The current method for detecting the biomarkers requires the use of a spectrophotometer. While the spectrophotometer serves as an accurate measurement technique, it is ill-suited for being able to process patient samples on-site due to its large physical size, delicate parts and complexity in operation. As an alternative, a low cost, single-use screen-printed iridium/carbon biosensor is proposed. This sensor is able to oxidize both NADH and hydrogen peroxide at the potential +0.27V versus the printed Ag/AgCl reference electrode. Using these sensors, and immobilized enzymes, the specific concentrations of total bile acid and adenosine deaminase could be determined amperometrically. Variables including: pH, temperature and testing media were explored to understand their influence on the current output as related to initial biomarker concentration. Testing results show that the constructed sensors can accurately, and repeatedly, detect the presence of the biomarkers total bile acid and adenosine deaminase. It is believed that these initial results show the sensor to be a promising viable method of detecting liver disease, particularly in those individuals who currently have limited access to medical diagnosis.

1. Introduction

The impact of liver disease can be felt world-wide. For the estimated over 170 million people suffering from liver disease, such as hepatitis and cirrhosis, it is essential that they are diagnosed early on [1]. Quality of life and effectiveness of treatment depend heavily on the how soon one is diagnosed. Unfortunately, those areas of the world that are suffering the most are also the areas that can least afford medical diagnosis. Hepatitis infection serves as an example. In Egypt, Bolivia, Tanzania, and Mongolia it is estimated that more than 10% of the population is infected [2]. While in China and Brazil infection rate ranges between 2.5 to 5% of the population [2]. Several of these areas have populations that are economically impoverished or live in remote areas.

Traditional methods of diagnosis rely on the need of physically large, costly and non-portable equipment such as spectrophotometers. While accurate, these techniques do not allow for the rapid processing of samples on site and need specialized instruction to operate. What is needed is an easy, portable, accurate method of diagnosing liver disease. A method such as this would allow medical workers to quickly identify those patients at risk. Once identified, the proper care could be coordinated. This would allow earlier detection, helping to reduce the spread of liver disease and allow treatment to be more effective.

In recent years there has been a growing interest in the field of biosensors. Biosensors traditionally utilize biologically derived recognition elements to produce electrochemically detectable species. Using electrochemical methods, the concentration of the species can be detected. The biosensors are designed to detect specific biomarkers that are characteristic of certain conditions in the body. Two specific biomarkers that have shown clinical significance in detecting liver disease are total bile acid (TBA) and adenosine deaminase (ADA). TBA is the collective term given for the six bile acids found in the body. Of those six two acids, cholic and chenodeoxycholic acid, comprise more than 85% of the total amount. They are produced by the liver and stored in the gallbladder for use in digestion of fatty foods. Comparison studies between healthy individuals and those with liver disease have shown that elevated TBA levels correlate very closely with the presence, and to some extent severity, of liver disease [3, 4]. An additional equally useful biomarker adenosine deaminase (ADA) has also been shown to be clinically relevant. ADA is an enzyme that is produced by the body as an immunological response to an infection. The enzyme helps to catabolize adeno-containing compounds. Clinical research has shown that elevation of ADA activity in an individual indicates the presence of liver infection. Studies have also shown that careful monitoring of ADA activity during treatment is a good predictor of the effectiveness of the treatment [5-7].

Our approach to detecting these biomarkers is to use a platform sensor technology that has been developed in house. As shown in Figure 1, the sensor is fabricated using thick-film screen printing methods. Using this method allows for flexibility of design while keeping the per sensor cost low.

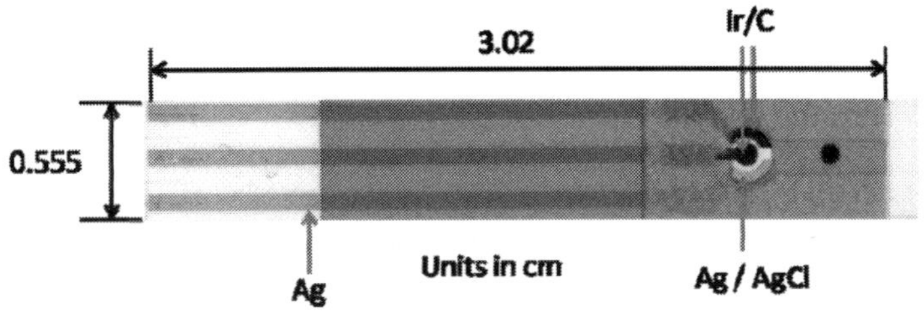

Figure 1. Thick-film, screen printed platform sensor.

The sensor is in a three electrode configuration with 5% iridium deposited on carbon serving has the working and counter electrode, while using a printed Ag/AgCl as the reference electrode. Printed silver ink serves as the connecting wires for the system. The detection schemes for each of the biomarkers are as follows:

ADA:

$$Adenosine + H_2O \xrightarrow{adenosine\ deaminase} Inosine + NH_3 \tag{1.1}$$

$$Inosine + phosphate \xrightarrow{PNP} Hypoxanthine + Ribose - phosphate \tag{1.2}$$

$$Hypoxanthine + 2H_2O + O_2 \xrightarrow{XO} Uric\ Acid + 2H_2O_2 \tag{1.3}$$

TBA:

$$Bile\ Acid + NAD^+ \xrightarrow{3\alpha-HSD} NADH + Oxidized\ Bile\ Acid \tag{1.4}$$

For the detection of ADA, the electrochemically detectable chemical H_2O_2 is produced. While for TBA detection NADH serves as the electrochemically detectable species. Attempts have been made in the past to detect both biomarkers but the results have been limited. In research by Alberry, efforts were focused on using an organic based salt as the catalyst for detection of NADH. Because of the high oxidation potentials needed for detection the sensor system required the use of size selective membranes to filter out potential interfering species such as ascorbic and uric acid. In work by Koide the use of a filtering membrane was still required but an additional detection enzyme was used to produce H_2O_2 [8]. While the oxidation potential was lower than Albery's studies, +0.5V versus an Ag/AgCl reference electrode was still required [9]. The advantages of the proposed sensor include the direct electrochemical detection of NADH at a low overpotential, +0.27V. This allows for a less complex system that is less prone to oxidizing interfering species. There are few studies that have focused on the detection of ADA. Fortunately, there is research available on the detection of the intermediate species hypoxanthine, inosine and purine. Work by Carsol and Okuma focused on the detection of hypoxanthine and inosine as a method of detecting fish freshness [10, 11]. While the sensors were able to detect the species, the systems utilized involved large flow columns and required large sample volumes. This was acceptable because of the source of the samples, fish for consumption, but would not be suitable to the application that we are focused on. In work by Llaudet, sensors were developed that allowed for the detection of purine within the brains of rats[12]. Design of the sensor required metal filaments to be coated with immobilized enzymes. Using a potential of +0.5V, versus an Ag/AgCl reference electrode, the purine levels could be monitored. Showing good utility for detection of purine, this sensor still has a relatively high operating potential that required the use of a selective polymer for operation. The sensor also would not be feasible for production on a scale necessary to meet the expected large numbers of patients with liver disease.

In these studies we used the above described thick-film, screen-printed disposable sensor technology to construct two sensors. In the first sensor we utilized the enzyme 3α-hydroxysteroid dehydrogenase (3α-HSD) to enzymatically form NADH from provided NAD^+.

The current generated by the oxidation of NADH was monitored amperometrically at +0.27V, versus Ag/AgCl reference electrode. The current is related to the initial concentration of bile acid in solution. The second sensor was developed to measure the enzymatic activity of adenosine deaminase in solution. Detection relied upon the immobilized enzymes purine nucleoside phosphorylase and xanthine oxidase to produce the electrochemically active chemical H_2O_2. The sensor was able to detect the H_2O_2 amperometrically at a potential of +0.27V and related to the initial concentration of ADA in solution. Behavior of the sensors was first examined in buffer to understand their behavior and to optimize operating parameters. Next, limits of the sensors were explored by adjusting operating parameters such as enzyme concentration, pH and species interaction. Finally the sensors were tested in the physiological solutions of blood and serum to confirm their usefulness. The results of the work indicate that both sensors are capable of detecting their corresponding biomarker with good linearity over the studied range. The results prove them to be promising for future development as a key tool in the fight against liver disease.

2. Experimental

2.1 Materials and reagents

Lyophilized xanthine oxidase (EC 1.1.3.22), purine nucleoside phosphorylase (EC 2.4.2.1), 3α-hydroxysteroid dehydrogenase (EC1.1.1.50) powder and adenosine deaminase (EC 3.5.4.4) in 50% glycerol solution were purchased from Diazyme Laboratories (San Diego, CA, USA). Ir/C particles (5% Ir) were purchased from BASF (Somerset, NJ, USA). Potassium chloride, 3% hydrogen peroxide solution, sodium phosphate monobasic, sodium phosphate dibasic heptahydrate, sodium phosphate tribasic, NADH and NAD^+ were purchase from Fisher Scientific (Hampton, NH, USA). Adenosine, sodium cholate, taurocholic acid – sodium salt and sodium taurochenodeoxycholate were purchase from Sigma-Aldrich (St. Louis, MO). Bovine serum used was from Invitrogen (Carlsbad, CA). The murine blood and plasma were provided from the Case Western Reserve University Cystic Fibrosis Center Animal Core. Additional chemicals used were of analytical grade. Buffers and solutions made were prepared using de-ionized water.

2.2 Experimental sensor fabrication

The formulation of the Ir/C ink used for the working and the counter electrode has been described in previous work[13]. Using a polyester substrate, silver ink was printed to serve as the electrical contacts. As mentioned screen-printed Ir/C ink served as the working and counter electrodes. A printed Ag/AgCl served as the reference electrode during testing. The geometric area of the printed electrode using in testing was 7.85×10^{-3} cm^2. This construction served as the platform for both the TBA sensor and the ADA sensor.

2.3 Assembly of TBA and ADA sensor

Using the platform sensor described above, a series of enzymes were immobilized onto the surface to serve as the bio-recognition elements. To construct the TBA sensor, 0.05

units of 3α-HSD was deposited onto each sensor. To do this, 0.7mg of lyophilized 3α-HSD was dissolved into 737 μL of pH 8.9 100mM phosphate buffer containing 150 mM potassium chloride and 100mM NAD^+. After mixing and storing the solution at 4°C for 10 minutes, 1μL of the solution was pipetted onto the working electrode of the sensor prototype. Once complete, the sensor was allowed to dry at 4°C for one hour before use. The construction of the ADA sensor was performed in a similar manner. Using 110 μL of pH 7.5 100mM phosphate buffer containing 150mM KCl, 12 mg of xanthine oxidase and 1.4 mg of purine nucleoside phosphorylase were dissolved. After cooling for 5 minutes at 4°C, 1 μL of solution was pipetted onto the working electrode surface and allowed to dry at 4°C. The result was 1.66 units of xanthine oxidase and 0.32U of purine nucleoside phosphorylase per sensor. Operation of the TBA sample combined a 3 μL bile acid containing sample with a pH 8.9 buffer solution to adjust the operating pH. A 6 μL droplet was placed on the sensor and testing was initiated. ADA sensor operation combined a 3 μL solution of ADA containing solution with a buffer solution containing 10.3 mM of adenosine. 3 μL of the combined sampled was applied to the sensor and testing was initiated.

3. Results

As shown in equations 1.3 and 1.4, the final enzymatic products of the two sensors were NADH and H_2O_2. In order to assess the feasibility of using the sensor for detection of these chemical phosphate buffer solutions were spiked with known concentrations of each chemical and cyclic-voltammetry studies were conducted. Figure 2 shows results for a system with, and without 200 μM of NADH and a separate system with, and without 200 μM of H_2O_2.

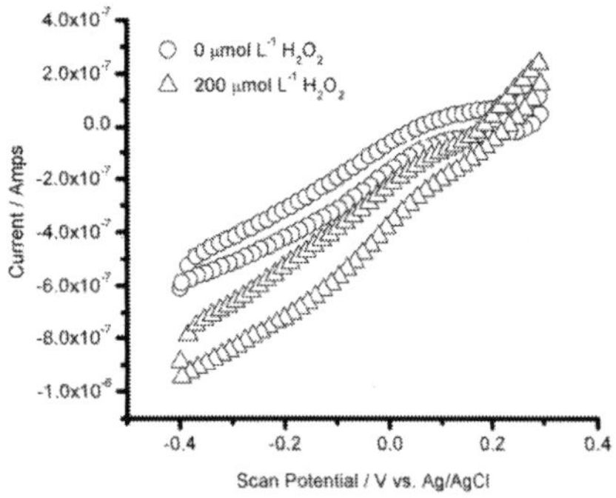

Cyclic-voltammetry scans with and without H_2O_2 in a 0.1M pH 7.0 phosphate buffer with 150 mM KCl

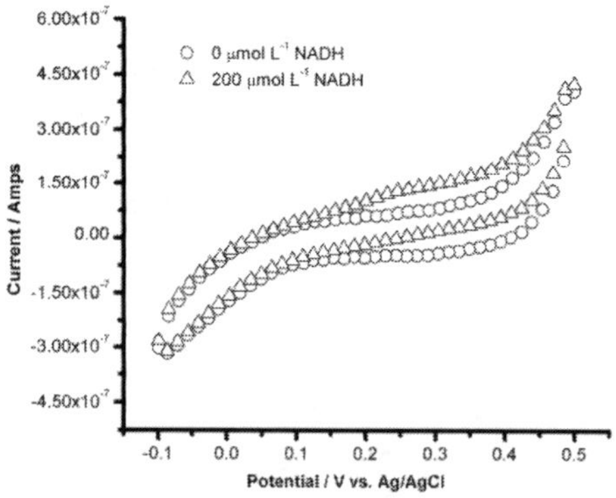

Cyclic-voltammetry scans with and without NADH in a 0.1M pH 7.0 phosphate buffer with 150 mM KCl

Figure 2. Cyclic-voltammetric scans of target species in pH 7.0 phosphate buffer

Figure 2 shows that there is a clear difference in the response current for those solutions containing the target species, NADH and H_2O_2, and those without. From this data we were able to select an oxidation potential of +0.27V, versus an Ag/AgCl reference electrode, as the operating potential for the amperometric test. The choice of the potentials allowed for good sensor response but was low enough to minimize oxidation of possible interfering species such as uric and ascorbic acid. Using the chosen potential, amperometric studies were next conducted to gain an appreciation of the linearity of response to increasing concentrations of NADH and H_2O_2. In figure 3, concentration versus current response at the same time point is shown for the two species.

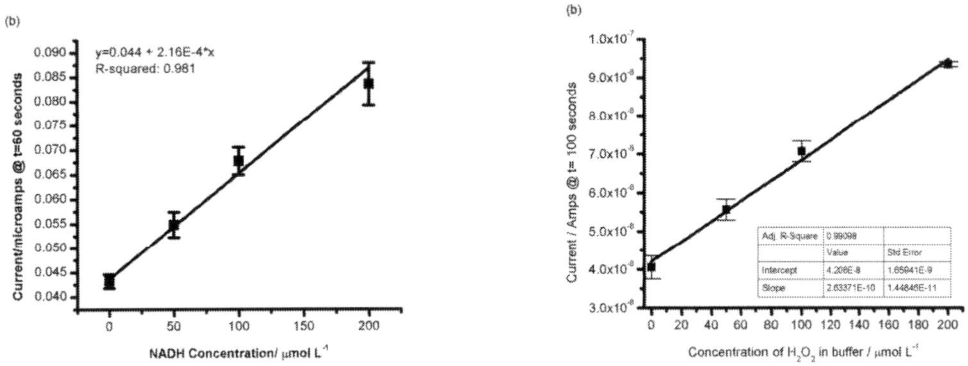

Amperometric response of sensor at 60 seconds in pH 7.0 phosphate buffer with known NADH additions

Amperometric response of sensor at 100 seconds in pH 7.0 phosphate buffer with known H_2O_2 additions

Figure 3. Amperometric response of the sensor in pH 7.0 phosphate buffer

The observed response in both cases was found to be well described by a linear relationship. From this we learned that the sensors can accurately detection NADH and H_2O_2 over the anticipated detection range with a good linear response at the low potential of +0.27V, versus the Ag/AgCl reference electrode. This gave confidence that if the sensor is properly constructed that we would be able to monitor our biomarker.

The next portion of the studies focused on the response of a fully constructed sensor in phosphate buffer solution. Shown in figure 4 is the response of the two sensor designs to their respective biomarker.

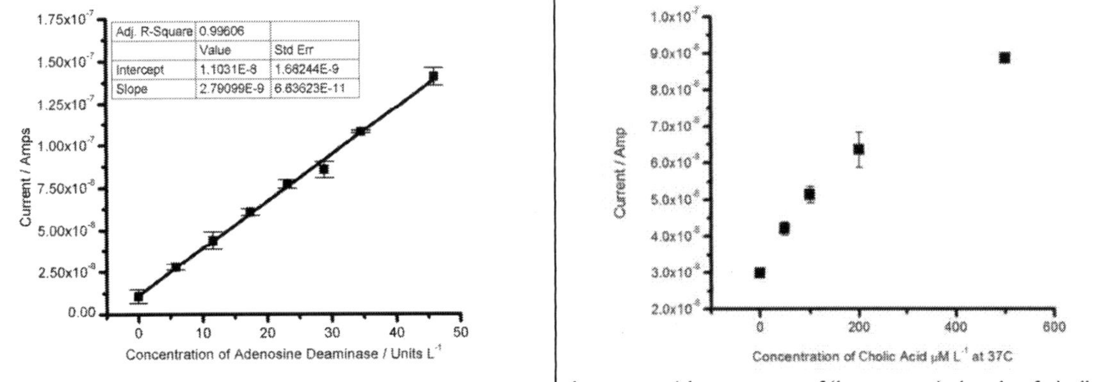

Amperometric response of the sensor to levels of ADA in phosphate buffer

Amperometric response of the sensor to levels of cholic acid in phosphate buffer

Figure 4. Amperometric response of the biosensor to their corresponding biomarker

In the case of the TBA sensor, figure 4 shows the response of cholic acid. The anticipated range of the sensor is 0 to 200 µM, but was tested at 500 µM to understand the limitation of the sensor. It was observed that the sensor has a linear response over the target range. Change in the slope was observed at the upper concentration, 500 µM, but this concentration is well beyond the anticipated range. A more extensive review of the performance of the sensor to bile acid combinations can be found in another source [13]. Response of the ADA sensor over the target range of 0 to 48 U/L also showed good linear response. Additional studies were conducted to confirm that the sensor was detecting the enzymatic response of the ADA. It was found that this was the case and was not a potential intermediate that is being detected.

Finally both sensors were tested in a physiological fluid more similar to what would be expected as the final testing medium. Figure 5 shows the sensor response of both systems to their respective biomarker in those mediums.

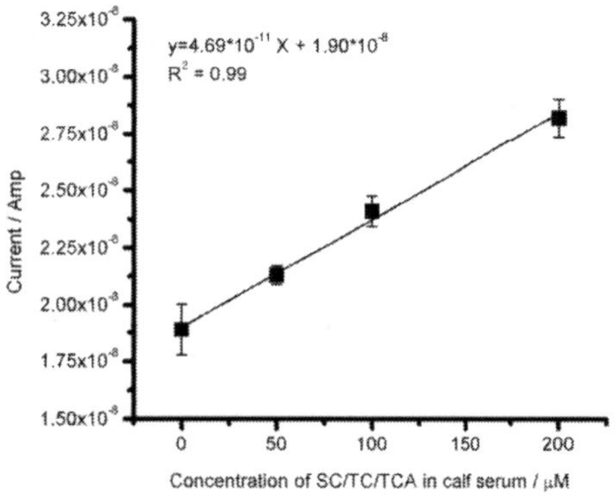

Amperometric response of the biosensor in calf serum to a combination of three bile acids

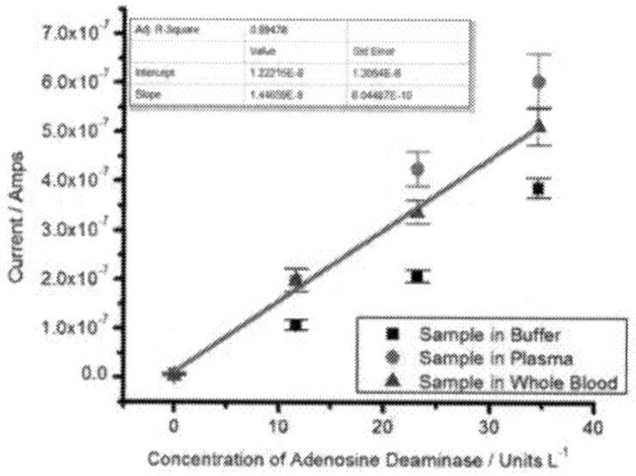

Amperometric response of the ADA biosensor in buffer, mouse whole blood and mouse plasma

Figure 5. Amperometric response of both biosensors in physiological similar testing fluids

The response of the biosensors to their respective biomarker in a physiological similar fluid was good. In the case of the TBA sensor, the response showed good linear performance over the entire range with low variation. The three acids added to the system were in relative proportion to those found in the body. For the ADA sensor, good linear response was seen in all three fluids. Although in the case of plasma and whole blood the current response was elevated beyond that of buffer. The fact that the sensor response for a zero concentration is the same for all three fluids indicated that the elevated current for higher concentrations was not due to impurities.

4. Conclusions

Liver diseases, such as hepatitis and cirrhosis, affect millions of people every year. Significant hurdles in diagnosing potentially infected individuals include the lack of available funds and the remote locations that testing needs to be conducted in. By providing disposable, screen-printed biosensors that are capable of measuring key liver disease biomarkers it is believed a difference can be made in clinical efforts to help patients. We have developed two biosensors, based on the detection of total bile acid and adenosine deaminase activity, that we feel are suitable for this problem. Advantages of the sensors are the ability to detect NADH and H_2O_2 at a low operating potential of +0.27V using small volumes of fluids. The results show the sensors to have a linear response of the target biomarker range. We also showed the sensors to have good performance in buffer and more physiological similar fluids such as serum, plasma and blood. It is believed these results provide a good foundation for further efforts to produce a sensor for this much needed application.

5. References

[1] B. D. Lindenbach, M. J. Evans, A. J. Syder, B. Wolk, T. L. Tellinghuisen, C. C. Liu, T. Maruyama, R. O. Hynes, D. R. Burton, J. A. McKeating, and C. M. Rice, "Complete Replication of Hepatitis C Virus in Cell Culture," *Science,* vol. 309, no. 5734, pp. 623-626, 2005.

[2] W. H. O. (WHO), "Global prevalence of hepatitis A, B, and C," *Weekly Epidemiological Record,* vol. 77, no. 6, pp. 41-48, 2002.

[3] S. Fischer, U. Beuers, U. Spengler, F. M. Zwiebel, and H.-G. Koebe, "Hepatic levels of bile acids in end-stage chronic cholestatic liver disease," *Clinica Chimica Acta,* vol. 251, no. 2, pp. 173-186, 1996.

[4] F. Jorquera, M. J. Monte, J. Guerra, S. Sanchez-Campos, J. A. Merayo, J. L. OlcÃ³z, J. GonzÃ¡lez-Gallego, and J. J. G. Marin, "Usefulness of combined measurement of serum bile acids and ferritin as additional prognostic markers to predict failure to reach sustained response to antiviral treatment in chronic hepatitis C," *Journal of Gastroenterology & Hepatology,* vol. 20, no. 4, pp. 547-554, 2005.

[5] S. Kaya, E. S. Cetin, B. C. Aridogan, S. Arikan, and M. Demirci, "Adenosine deaminase activity in serum of patients with hepatitis -- a useful tool in monitoring clinical status," *J Microbiol Immunol Infect,* vol. 40, no. 4, pp. 288-92, Aug, 2007.

[6] A. Kalkan, V. Bulut, O. Erel, S. Avci, and N. K. Bingol, "Adenosine deaminase and guanosine deaminase activities in sera of patients with viral hepatitis," *Mem Inst Oswaldo Cruz,* vol. 94, no. 3, pp. 383-6, May-Jun, 1999.

[7] F. Kobayashi, T. Ikeda, F. Marumo, and C. Sato, "Adenosine deaminase isoenzymes in liver disease," *Am J Gastroenterol,* vol. 88, no. 2, pp. 266-71, Feb, 1993.

[8] S. Koide, N. Ito, and I. Karube, "Development of a micro-planar amperometric bile acid biosensor for urinalysis," *Biosensors & Bioelectronics,* vol. 22, no. 9/10, pp. 2079-2085, 2007.

[9] W. J. Albery, R. B. Lennox, E. Magner, G. Rao, D. Armstrong, R. Hermon Dowling, and G. M. Murphy, "An amperometric enzyme electrode for bile acids," *Analytica Chimica Acta,* vol. 281, no. 3, pp. 655-661, 1993.

[10] M. A. Carsol, and M. Mascini, "Development of a system with enzyme reactors for the determination of fish freshness," *Talanta,* vol. 47, no. 2, pp. 335-342, 1998.

[11] H. Okuma, and E. Watanabe, "Flow system for fish freshness determination based on double multi-enzyme reactor electrodes," *Biosensors and Bioelectronics,* vol. 17, no. 5, pp. 367-372, 2002.

[12] E. Llaudet, N. P. Botting, J. A. Crayston, and N. Dale, "A three-enzyme microelectrode sensor for detecting purine release from central nervous system," *Biosensors and Bioelectronics,* vol. 18, no. 1, pp. 43-52, 2003.

[13] B. Bartling, L. Li, and C.-C. Liu, "Determination of total bile acid levels using a thick-film screen-printed Ir/C sensor for the detection of liver disease," *Analyst,* vol. 134, no. 5, pp. 973-979, 2009.

Silicon Based Microfabricated Sensor for Detection of Organophosphate Compounds

Maryam Sayyah, Department of Chemical and Biomolecular Engineering, University of Illinois at Urbana-Champaign, Urbana, IL

Amin Salehi-Khojin, Department of Chemical and Biomolecular Engineering, University of Illinois at Urbana-Champaign, Urbana, IL

Kevin Y. Lin, Department of Chemical and Biomolecular Engineering, University of Illinois at Urbana-Champaign, Urbana, IL

Chelsea N. Monty, Department of Chemical and Biomolecular Engineering, University of Illinois at Urbana-Champaign, Urbana, IL

Richard I. Masel, Department of Chemical and Biomolecular Engineering, University of Illinois at Urbana-Champaign, Urbana, IL

In the present work, we utilize a microscale gas-liquid interface for use in a selective gas micro-detector for detection of toxic organophosphates (OP) compounds. Previous work in our lab has demonstrated that OP compounds can be selectively detected using such a sensor fabricated in polycarbonate; however, the sensor material is not inert and cannot be integrated with other MEMS-based silicon devices. In this work we focus on the design of a MEMS-based silicon sensor using both experiments and COMSOL simulations. A Teflon nano-porous membrane is used to enhance the stability of gas-liquid interface as well as sensitivity of detection. A toxic gas of interest is injected into the vapor microchannel and reacts with an alkaline oxime solution as it dissolves in the liquid phase. In the reaction cyanide ions are produced and detected using a gold electrode on the nano-porous membrane. Response is measured as change in open circuit potential between the working and reference electrodes integrated on a single chip. Due to the toxicity of OP compounds, an OP simulant (e.g. acetic anhydride) which undergoes a similar reaction mechanism has been used. The detection limit of this sensor design is in the parts per trillion levels or approximately 3×10^9 molecules.

In order to investigate the influences of important geometric parameters on detector performance, a finite element based commercial software, COMSOL 3.4 (Stockholm, Sweden) has been utilized. A 2D simulation of the system consists of gas and liquid micro-channels with a nanoporous gas-liquid interface. Coupled steady state Navier-Stokes, continuity and unsteady state mass transfer equations with chemical reaction have been numerically solved to establish a realistic model of the system. Simulation results indicate that using a nano-porous gas-liquid interface tremendously reduces diffusion time of cyanide ions, leading to a fast response of the detector compared to micron size membrane. Furthermore, it has been shown that the liquid channel depth and nano-membrane porosity are amongst the main parameters affecting the micro-sensor performance.

Experimental and simulation results demonstrate that the silicon based micro-detector proposed in this work can be a promising way to selectively detect ultra low levels of hazardous materials.

See more of this Session: Micro and Nanofabricated Sensors. In Honor of CC Liu I
See more of this Group/Topical: Topical 9: Sensors

Nanoelectrode Sensor Devices

Brian G. Willis, Chemical, Materials, and Biomolecular Engineering, University of Connecticut, Storrs, CT

A fundamental interest in Nanotechnology is the potential to take advantage of the unique properties of matter at the nanoscale, including the importance of electron tunneling. We present a novel approach to nanosensors based on tunneling spectroscopy in nanofabricated nanoelectrode devices. Using advanced reaction engineering capabilities based on selective area atomic layer deposition with in-situ electrical feedback, we prepare nanoelectrode junctions that are sensitive to molecular adsorption down to the single molecule level. Real time adsorption data are demonstrated by monitoring the tunneling current during exposure to a vapor sample. The electrical data show step-like features that are associated with individual molecular adsorption events. Molecular transport based on electric field enhanced surface diffusion is used to analyze the adsorption data and successfully captures the experimental trends. Inelastic electron tunneling spectroscopy is used to characterize the vibrational frequencies of the trapped molecule, and the successful spectroscopy is confirmed using isotopic substitution. Reversible adsorption/desorption and spectroscopy are demonstrated, showing the essential characteristics of an ideal sensor. The extension to massively parallel device arrays is demonstrated by adsorption in parallel devices. Lastly, the composition and structure of the nanoelectrodes are characterized and challenges for integration of the nanoelectrode devices into practical sensors are discussed.

See more of this Session: Micro and Nanofabricated Sensors. In Honor of CC Liu I
See more of this Group/Topical: Topical 9: Sensors

Thermal Analysis of SiC Microhotplates and Gas Sensors

Kane J. Miller, Chemical Engineering, University of Louisville, Louisville, KY

Xiao-an Fu, University of Lousiville, Lousiville, KY

Chemical microsensors using microhotplates have attracted a great interest in recent years because of tremendous demand and growth in the chemical sensor market. The demand for chemical microsensors comes from new legal regulations on environmental air pollutions such as automobile exhausting gases NOx, CO, and increasing health awareness in the population of industrialized countries. Microfabricated gas sensors have advantages of micro size, fast response, and low power consumption. Microhotplate and metal oxide as sensing element are key factors for practical applications of microhotplate-based gas sensors. Current metal and polysilicon based microhotplates can not reach a temperature above 500 oC. Oxidation of metal and polysilicon heating elements at high temperatures also induces increasing power consumption and reliability problems.

SiC is well known for its combination of excellent mechanical, electrical, and chemical properties, which include high mechanical strength, wide bandgap, high breakdown voltage, high thermal conductivity, resistance to oxidation and creep, and chemical inertness to oxidation. It is a very promising compound semiconductor for MicroElectroMechanical Systems (MEMS) and microelectronics for high temperature and extremely harsh environment applications.

This work reports our effort to use MEMS software CoventorWare to do thermal simulation of silicon carbide (SiC) microhotplates to guide the design of SiC microhotplate based gas sensors for harsh environment (high temperature, high corrosion, high oxidation) applications. SiC has excellent oxidation resistance at high temperatures comparing with metal and polysilicon. Heavily doped SiC thin films deposited by low pressure chemical vapor deposition (LPCVD) are used for microhotplate heating element. Multiple simulations have been done to explore the effect of microhotplate geometry and supporting membrane on the maximum temperature and the temperature distribution. The preliminary experiment and simulation results indicate that SiC microhotplate can reach a temperature above 600 oC. Further simulation will be performed to determine the effect of changing the insulating material from silicon nitride to, for example, silicon dioxide, as well as altering the overall size of the hotplate, the thicknesses of each layer, and the doping concentration of the silicon carbide. The collective purpose of these simulations will be to optimize the power consumption maximum temperature and the temperature uniformity across the hotplate, making it ideal for use in gas sensors. Microfabrication of SiC microhotplates and gas sensors for sensing CO, NOx, and NH3 are in progress and will be characterized. The tested results of the SiC microhotplates will be compared with the simulated results and the characterization of gas sensors will also be reported.

See more of this Session: Micro and Nanofabricated Sensors. In Honor of CC Liu I
See more of this Group/Topical: Topical 9: Sensors

Polymer-Functionalized Aligned Carbon Nanotubes for Biological and Chemical Sensing

Liming Dai
Department of Chemical and Materials Engineering, School of Engineering and UDRI, University of Dayton, 300 College Park, Dayton, OH 45469-0240, USA
Email: ldai@udayton.edu

We have immobilized glucose oxidase (GOX) onto vertically-aligned carbon nanotube arrays by electropolymerization of pyrrole in the presence of GOX as new sensing active materials for making advanced glucose sensors with a high sensitivity. Single-strand DNA chains have also been chemically grafted onto the aligned carbon nanotube electrodes, leading to novel aligned carbon nanotube-DNA sensors of a high sensitivity and selectivity. Furthermore, we have recently developed new thin-film-type chemical vapour sensors by partially coating vertically-aligned carbon nanotube arrays with an appropriate polymer thin film along their tube length. Subsequent absorption and desorption of chemical vapors by the polymer matrix could cause changes in the inter-tube distance, and hence the surface resistance across the nanotube film. Simple measurements of the resistance change, therefore, constituent the nanotube-polymer chemical vapor sensors. These rationally-designed aligned carbon nanotube-polymer composite films are flexible and can be effectively integrated into many systems for a wide rang of potential applications, including their use as multifunctional sensors for sensing chemical vapors, mechanical deformations, thermal and optical exposures. In this talk, I will present some of our work on aligned carbon nanotubes to provide rational concepts for the design and development of polymer-functionalized aligned carbon nanotube arrays for biological and chemical sensing applications.

Invited 4

Heidi B. Martin, Case Western Reserve University, Cleveland, OH

To be added

See more of this Session: Micro and Nanofabricated Sensors. in Honor of CC Liu II
See more of this Group/Topical: Topical 9: Sensors

A Multi-Electrode Array Microbiosensor for Monitoring of Neurotransmitter Concentration Changes In Vivo

Vanessa M. Tolosa, Chemical and Biomolecular Engineering, University of California, Los Angeles, Los Angeles, CA

Kate M. Wassum, Psychiatry and Biobehavioral Sciences, University of California, Los Angeles, Los Angeles, CA

Nigel T. Maidment, Psychiatry and Biobehavioral Sciences, University of California, Los Angeles, Los Angeles, CA

Harold G. Monbouquette, Department of Chemical and Biomolecular Engineering, University of California, Los Angeles, Los Angeles, CA

We are designing, producing, and implementing micromachined, multielectrode microprobes for the detection of neurotransmitters in near real-time in the brains of live rodents. Over 100 devices are constructed simultaneously on four-inch silicon wafers with two to five, micron-sized electrodes per probe using micro-electro-mechanical-systems (MEMS) fabrication technologies. One or more microelectrode sites per probe is chemically modified with permselective polymer films and immobilized glutamate oxidase for the electroenzymatic detection of glutamate. Another site is modified for use as an on-probe reference electrode. Glutamate is detected selectively with sub-second response time in the presence of the electroactive interferents, dopamine and ascorbic acid. In the anesthetized rat, these glutamate biosensors detect cortically-evoked glutamate release in the ventral striatum. In the freely moving rat, stress-induced glutamate release is detected in the dorsal striatum. Currently, these microprobes are being explored for use in the study of drug-seeking behavior in rodents. The microelectrode microsensors also may serve as a platform for the creation of multi-analyte (e.g., glutamate and dopamine) sensing devices useful for measurements in vivo.

See more of this Session: Micro and Nanofabricated Sensors, in Honor of CC Liu II
See more of this Group/Topical: Topical 9: Sensors

A Novel Fabrication Method for Producing Arrays of Nanopore Devices for Biomolecule Analysis

Amir A. Ahmadi, Chemical and Biomolecular Engineering, Georgia Institute of Technology, Atlanta, GA

Sankar Nair, School of Chemical & Biomolecular Engineering, Georgia Institute of Technology, Atlanta, GA

Nanopore-based single-molecule analysis of biomolecules such as DNA and proteins is a subject of strong scientific and technological interest. In this work, a novel fabrication procedure using traditional microelectronics tools is developed to produce arrays of finely tunable nanopores in free-standing silicon nitride across a wafer. The process couples the high resolution and scalability of electron beam lithography and atomic layer deposition. This overcomes the limitations of other methods used to produce pores, such as transmission electron beam (scalability, lack of automation) and focused ion beam (reproducibility). Once nanopores of order 10 nm or less are achieved, DNA translocation will be tested in individual devices using optimized electrical stimuli, with the goal of approaching ultra-rapid high resolution size separation of strands.

See more of this Session: Micro and Nanofabricated Sensors, in Honor of CC Liu II
See more of this Group/Topical: Topical 9: Sensors

Ultrasensitive NH3 Gas Sensor From Nanostructured TiO2 -Polyaniline PN Heterojunctions

Jian Gong, School of Chemical and Biomolecular Engineering, Georgia Institute of Technology, Atlanta, GA

Yinhua Li, School of Chemical and Biomolecular Engineering, Georgia Institute of Technology, Atlanta, GA

Zeshan Hu, IPST at Georgia Tech, Georgia Institute of Technology, Atlanta, GA

Zhengzhi Zhou, School of Chemical and Biomolecular Engineering, Georgia Institute of Technology, Atlanta, GA

Yulin Deng, School of chemical and biomolecular engineering, Georgia Institute of Technology, Atlanta, GA

Increasing demands for ever more sensitive chemical sensors have led to an upsurge of interest in combining nanostructured metal oxides with polymers. The PN junction materials consisting of semiconductive inorganic oxides and conductive polymers have been widely used to enhance the sensitivity of nanostructured sensors. Most of the enhancements on the sensing sensitivity are due to a synergy of organic and inorganic parts. Although sensors based on nanostructured PN junction have been reported, there are still many challenges in making a ultrasensitive sensor. Field-effect transistor is the most common technique that uses PN junctions to enhance nanosensor sensitivity and selectivity. In this communication, we present an alternative way to take the advantages of PN junctions for fabricating ultrasensitive nanostructrued gas sensors. Specifically, nanoparticles of a P-type conductive polymer, polyaniline (PANI), are enchased on electrospun semiconductive TiO2 nanofiber surface. The PN heterojunctions formed between TiO2 nanofibers and PANI nanoparticles function as electric current switches when NH3 gas is absorbed by PANI nanoparticles. The PN heterojunctions formed between TiO2 nanofibers and PANI nanoparticles function as electric current switches when NH3 gas is absorbed by PANI nanoparticles. The sensor constructed with these unique PN heterojuctions can detect as low as 50 ppb of NH3.

See more of this Session: Micro and Nanofabricated Sensors, in Honor of CC Liu II
See more of this Group/Topical: Topical 9: Sensors

Microfluidic Pneumatic Logic Circuits and Digital Pneumatic Microprocessors

Minsoung Rhee[1] and Mark A. Burns[1,2,*]

[1]Department of Chemical Engineering, the University of Michigan
2300 Hayward St. 3074 H.H. Dow Building, Ann Arbor, MI 48109-2136
[2]Department of Biomedical Engineering, the University of Michigan
2200 Bonisteel Blvd, Ann Arbor, MI 48109-2099

Many microfluidic applications typically require the integration of a number of valves, pumps, and other active components, most of which require auxiliary off-chip control. Although recent advances in pneumatic microvalves [1-2] have enabled large-scale integration of microfluidic components to perform hundreds of operations in parallel by multiplexing control of embedded operational valves [1-4], the need for a large number of dedicated external control lines limits the practical use of integrated microfluidic systems. The complex valve control can be simplified by integrating microfluidic devices using on-chip fluidic logic networks [5]. Much like the development of electronic logic gates simplified the construction and operation of complex electronic devices, pneumatic logic gates could considerably reduce the number of required off-chip controllers. Various microfluidic logic gates operated on Boolean rules have been proposed to direct internal flows in complex networks and perform simple on-chip calculations [6-10]. Although conceptually powerful, most of the developed microfluidic logic gates rely on different input/output types and, therefore, the output cannot be used as an input to directly actuate subsequent logic gates. This non-cascadable nature inhibits further scaling and feedback routing for more complex circuits. Other approaches utilizing droplet-based input/output for logic gates exhibit excellent cascadability but construction of control platforms from these systems is difficult since the trajectories or existences of droplets cannot be used for actuation of active control components [9-10]. Most importantly, no parallel operations have been achieved from programmable serial input signals.

We have developed pneumatic logic circuits and microprocessors built with microfluidic channels and valves in polydimethylsiloxane (PDMS) that are capable of performing various combinational and sequential logic calculations with binary pneumatic signals (atmosphere and vacuum). The pneumatic logic circuits perform various combinational and sequential logic calculations with binary pneumatic signals (atmosphere and vacuum), producing cascadable outputs based on Boolean operations. A complex microprocessor is constructed from combinations of various logic circuits and

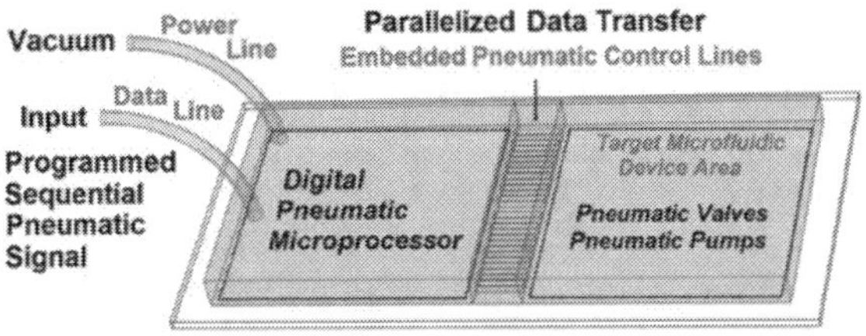

Fig. 1. Schematics of a digital pneumatic microprocessor

receives pneumatically encoded serial commands at a single input line. The device then decodes the temporal command sequence by spatial parallelization, computes necessary logic calculations between parallelized command bits, stores command information for signal transportation and maintenance, and finally executes the command for the target devices.

Microfluidic digital pneumatic circuits that operate on serially-encoded command signals are constructed from a number of different pneumatic components such as flip-flops, logic gates, and shift registers. Fig. 1 shows a schematic of the experimental setup with a pneumatic microprocessor platform for a target device that requires a number of individual control lines. The pneumatic circuits operate based on the pneumatic Boolean rules (0 and 1). Two distinct pressure levels are used for the binary information: atmospheric pressure (bit value = 0) and vacuum (bit value = 1) (Fig. 2a). Due to the membrane deflection at valves, any open valve chambers reflects while closed valves do not, providing a simple and straightforward indicator to visually examine the binary pressure levels (Bit 0 and Bit 1) (Fig. 2b). Using this simple valve, we were able to construct a wide variety of pneumatic logic components. The

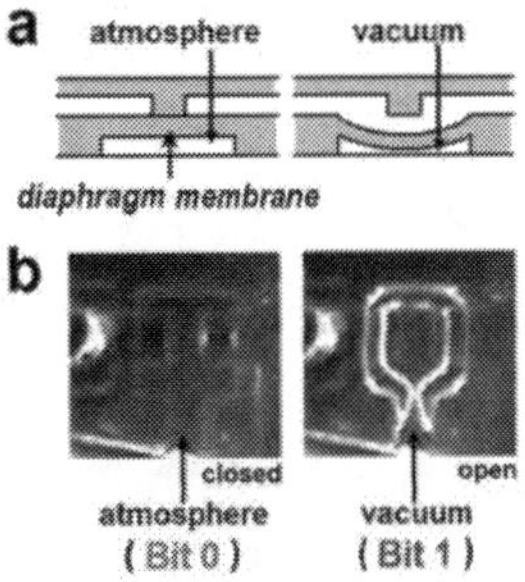

Fig. 2. Single pneumatic valve. (a) A thin diaphragm membrane opens in the presence of vacuum in the bottom chamber. (b) Explicit visualization of valve status by the amount of light reflection.

fundamental building unit in our system is the asymmetric microfluidic inverter or NOT gate (Fig. 3). It consists of a long microfluidic channel connected to vacuum and atmosphere at the ends, respectively, and employs a channel-separating microvalve that is normally closed at an atmospheric pressure but opens with a vacuum input [2]. While open holes to atmosphere allow for infinite supply of atmospheric pressure, vacuum can be also supplied to all circuits simultaneously from one large vacuum reservoir constructed in the third PDMS layer without separate individual tubing. On the analogy with electronics, this inverter is "powered" by the vacuum source connected to one end and "grounded" to atmosphere at the other end. Application of vacuum causes the input microvalve to open, resulting in a sharp pressure increase in the "output" chamber. For atmospheric input signals, the input valve is closed and the output chamber returns to a vacuum state. Note that the vacuum lines for all components can be supplied simultaneously from one large vacuum reservoir constructed in the third PDMS layer without separate individual tubing. Also, the output channel will be a dead end (e.g., a valve or a pump) even if it is an input to regulate other consecutive circuits. Figs. 3a and 3b show a PDMS asymmetric NOT gate connected to a valve on a target channel and operating at the two different binary pressure inputs. As long as the input pressure level is atmosphere (Bit 0), the valve in the NOT gate will remain closed and the output will remain at vacuum thus opening the valve on the target channel (Fig. 3a). Applying vacuum at the input opens the valve in the NOT gate and produces atmospheric pressure at the output, thus closing the valve on the target channel (Fig. 3b). Note that the opening and closing of the output valve can be visually observed, as stated before, by the change in light reflection. The operation of the valve can also be monitored by measuring the flow of gas or liquid through the target channel.

With simple additions to the asymmetric NOT gate, other logic gates can be constructed starting from the NOR gate and the NAND gate, i.e., two universal logic gates. The NOR gate is operated by two parallel input valves such that, when either one or both of the input valves open, air flows into the main channel and generates an atmospheric output signal. The output remains as vacuum only if both input valves are closed (i.e., both inputs are 0). Likewise, for the NAND gate, both of the two serial valves must be open together to produce an atmospheric output. Otherwise, vacuum will be maintained at the output. With these cascadable universal gates (i.e., NAND and NOR) along with the inverter, many other more complicated logic gates and circuits can be formed including AND, OR, XOR, and XNOR. Pneumatic bistable flip-flops, essential elements of serial data processing, were built from two regular inverters with each output fed backed to the other input (Fig. 4). For the two input valves, only one valve can be open at a time – the other must be closed. Once either valve is determined, both valves maintain their opposite statuses infinitely, remaining 'latched' in the absence of external perturbation. Interestingly, such PDMS flip-flops can be manually operated by pressing on the open valve. Due to the elasticity of PDMS, any open PDMS valve can be forced into the closed state by pressing on the top surface, even if the valve bottom chamber is under vacuum. Once one valve is forced closed, the other valve instantly opens and the whole structure maintain its new status even after the mechanical force is removed. Another mechanical 'touch' to close the newly opened valve will in turn make the original valve open again. This 'touch microfluidic control' thus would make possible complex operations of multiple active microfluidic components without any computerized control elements or complicated tubing.

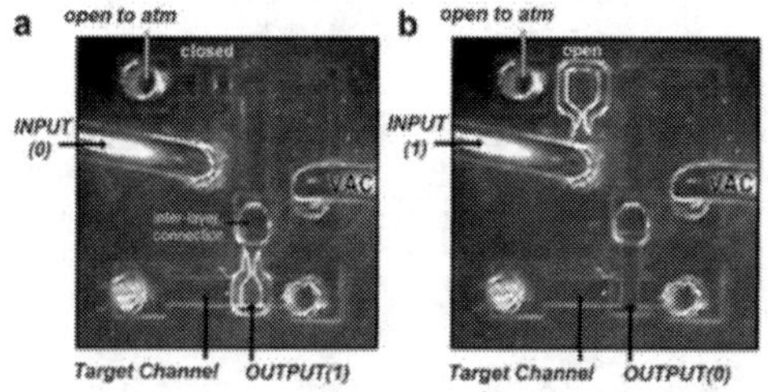

Fig. 3. (a) When the valve is closed (switch open), the main channel becomes discontinued, making vacuum-side channel a closed system. Vacuum will be transferred to the output without pressure drop. (b) when the valve is open (switch closed), the main channel becomes connected, generating pressure drop along the channel. Visual observations with the asymmetric NOT gate operating on the inputs of atmosphere (a) and vacuum (b).

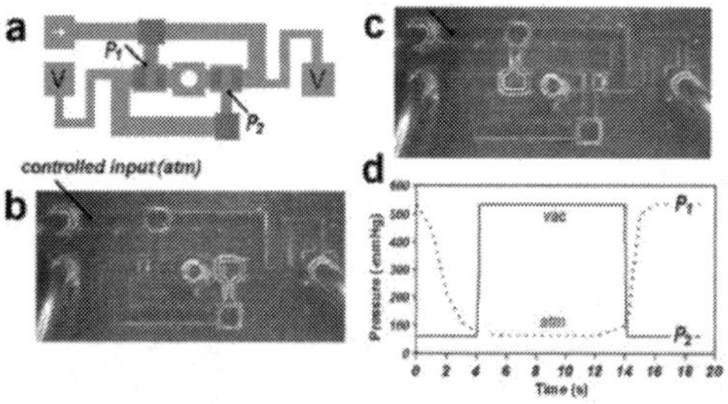

Fig. 4. Operation of a pneumatic bistable flip-flop. (a) CAD design of a flip-flop. The pressure levels of P_1 and P_2 are indicators for the flip-flop's states. (b) When input at P_1 of 0 is forcedly supplied, the input-side valve closes and the other valve instantly opens. (c) When the input of 1 is forcedly supplied, the input-side (left) valve opens and the other (right) valve instantly closes. (d) Plot of pressure level changes. P_1 (control) and P_2 (response) are always opposite to each other and change only by external perturbations. When the controlled pressure level (P_1) exceeds thresholds for the valve, instantaneous pressure changes in response (P_2) are observed.

Serial or sequential data is composed of a string of binary bits that are distinguishable temporally. Clock signals provide temporal placeholders for serial data where one clock cycle represents the smallest time that can carry a bit of information. The pneumatic clock signal consists of pulses oscillating at a certain frequency between atmospheric and vacuum pressure levels. Pneumatic D Latches with both inverted output and buffered output can be constructed from these bistable flip-flops and further extended to clocked D flip-flops. A shift register, critical for serial data processing, can decode the information contained in multiple sequential bits. By spatially separating temporal signals, shift registers can, essentially, parallelize serial information. A clocked

D Flip-flop consists of two consecutive D latches with a regular clock pulse supplied to the gate valve of the first D latch and an inverted clock pulse supplied to the gate valve of the second. Such flip-flops are then joined together to form multi-bit micropneumatic shift registers that perform serial-in parallel-out operations with pneumatic data. With n D flip-flops concatenated, $n+1$ bits of information are available at the parallel outputs. Fig. 5 shows a 4-bit pneumatic processor capable of parallelizing serial input data to control four individual valves independently to, in this case, mix four different color dyes. Such individually accessible valves were also used for multiplexing eight different flow channels using a 3-1 multiplexer. Instead of using separate trigger (command execution) signals to initiate signal transfer from shift registers to instruction registers, the processor can detect a termination code in the original serial input to execute commands. A pneumatic clock pulse generator was also embedded to eliminate the need for external solenoid clocking.

Such pneumatic microprocessors will function as a universal on-chip control platform to perform complex parallel operations for large-scale integrated microfluidic devices that require a roomful of control components. The miniaturization and on-chip integration of control elements has thus the great potential to significantly reduce the complexity of external controllers, interconnections and auxiliary equipment. Proposed digital pneumatic microprocessors can be a potent universal on-chip platform to autonomously manipulate microfluids in a high throughput manner. There is no theoretical limit to the number of target parallel outputs although physical limitations will certainly exist. Consequently, the pneumatic logic circuits and microprocessors are universal, operable with ease, and portable; they can thus significantly minimize cumbersome control, power consumption, and auxiliary external cost, making microfluidic analysis devices more accessible for various complicated applications, even at remote sites.

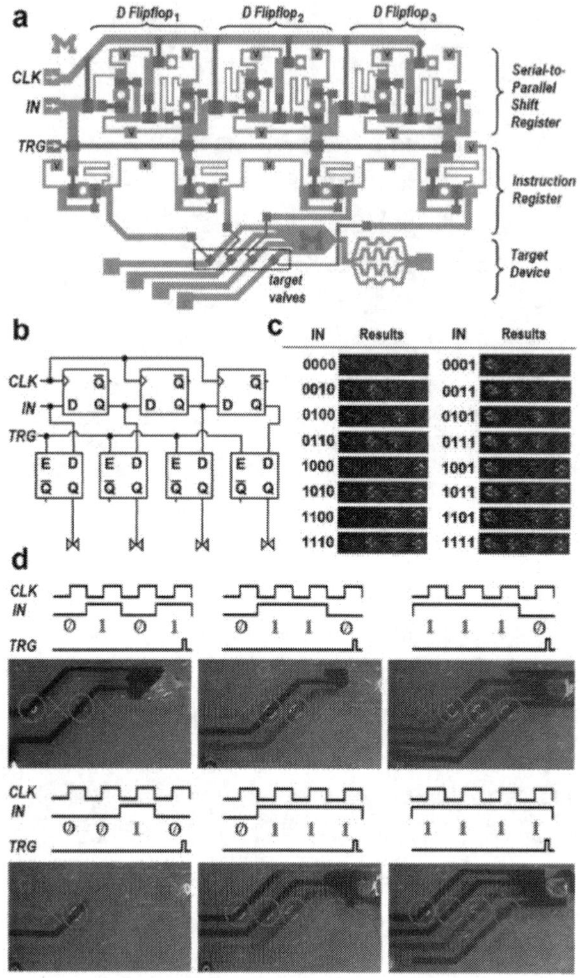

Fig. 5. 4-bit digital pneumatic microprocessor. (a) CAD design of the 4-bit processor consisting of a 3-bit MSR with three D flip-flops and a 4-bit instruction register that controls the target mixing device with four independent valves. Serially encoded input commands are shifted by accompanying clock pulses, generating four independent parallel outputs to control the corresponding command valves in the fluidic channels. (b) Electrical symbolic representation. (c) Statuses of four target command valves depending on the input commands. (d) Various command sequences were supplied to perform different mixing tasks. Note that the maintained outputs keep some valves open and the others closed during mixing operations.

References

1. M. A. Unger, H, -P. Chou, T. Thorsen, A. Scherer and S. R. Quake, *Science,* 2000, **288**, 113-116.
2. E. T. Lagally, J. R. Scherer, R. G. Blazej, N. M. Toriello, B. A. Diep, M. Ramchandani, G. F. Sensabaugh, L. W. Riley and R. A. Mathies, *Anal. Chem.*, 2004, **76**, 3162-3170.

3. T. Thorsen, S. J. Maerkl and S. R. Quake, *Science,* 2002, **298**, 580-584
4. W. H. Grover, A. M. Skelley, C. N. Liu, E. T. Lagally and R. A. Mathies, *Sens. Actuators B, Chem.*, 2003, **89**, 315-323.
5. W. H. Grover and R. A. Mathies, *J. Micromech. Microeng.*, 2007, **16**, 1378-1385.
6. T. Vestad, D. W. M. Marr and T. Munakata, *Appl. Phys. Lett.*, 2004, **84**, 5074-5075.
7. M. W. Toepke, V. V. Abhyankar and D. J. Beebe, *Lab Chip,* 2007, **7**, 1449-1453.
8. W. Zhan and R. M. Crooks, *J. Am. Chem. Soc.*, 2003, **125**, 9934-9935.
9. L. F. Cheow, L. Yobas and D. –L. Kwong, *Appl. Phys. Lett.,* 2007, **90**, 054107.
10. M. Prakash and N. Gershenfeld, *Science,* 2007, **315**, 832-835.

An Integrated Enzyme -Based Optical Glucose Sensor with Chemically Anchored in-Device Sensing Elements

Zhan Gao, Department of Chemical and Biological Engineering, Missouri University of Science and Technology, Rolla, MO

Chang-Soo Kim, Electrical & Computer Engineering, Missouri University of Science and Technology, Rolla, MO

David B. Henthorn, Department of Chemical & Biological Engineering, Missouri University of Science and Technology, Rolla, MO

An enzyme-based optical glucose microfluidic device is designed based on the quenching behavior of fluorophore, a ruthenium complex, towards oxygen molecules. Instead of traditionally used silicon- and glass-based material, a negative-tone high-aspect-ratio photoresist SU-8 was used as the main material of microfluidic channel structure. However, it is still a challenge to place the sensing elements before sealing the SU-8 cover layer because the temperature required for wafer bonding may cause enzyme denaturation. To avoid this, a novel in situ patterning method was introduced to covalently pattern enzyme containing hydrogel columns within the completed channels. The optimal glucose oxidase concentrations, ratio between enzyme and fluorophore loading, and the components of the materials used for fabricating the matrix were investigated. The characteristics of the sensor including the reversibility, reproducibility, and long-term stability were tested based on extensive experiment during a long time use. The principles underlying the sensing elements, like reaction kinetics of immobilized glucose oxidase, the diffusion of glucose into the PEG-DA matrix and the photobleaching of the fluorescent dyes were also studied. We anticipate that by using this technique, multiple sensing elements can be immobilized in these hydrogel microarrays within the microfluidic channels, offering a potential to create various novel biosensors for a wide range of applications.

See more of this Session: Micro and Nanofabricated Sensors. in Honor of CC Liu II
See more of this Group/Topical: Topical 9: Sensors

Stability and Reactivity of Enzymatic Papers

*Mohidus Samad Khan, Wei Shen and Gil Garnier**

Australian Pulp and Paper Institute (APPI), Department of Chemical Engineering
Monash University, Clayton, VIC 31800, Australia.
**Corresponding author: Gil.Garnier@eng.monash.edu.au*

Introduction

The high cost and the limited availability of tests and application devices often restrict the benefits of many discoveries in biotechnology, medicine and environmental science. There is a high need for low cost bioassays, in health and environmental diagnostics. Disposable materials can be very useful for medical or environmental purposes. Routine bioassays for the early detection of cancers and genetic conditions, for daily tests to monitor diabetes, and for instant water analyses of heavy metals and microbial activities are a few of the potential applications of disposable bioassays.

The successful commercialization requires bioassays and bioactive surfaces to be low cost, which can be achieved through high volume manufacturing and with commodity materials. Printing is a high speed technology, able to deliver patterns of functional fluids at an exceptional accuracy. Recent developments in bio-interface engineering have shown ink jet, a non-contact digital printing technique, as a promising method to create biomolecular patterns onto surfaces [1-4]. Ink jet can deposit biomolecules on porous substrates with high resolution: 20-40 μm and minimum cross contamination [5]. Ink jet has been used to print peptides, enzymes and cells on different substrates under laboratory conditions along with other printing techniques [6-14]. On the other hand, paper, a non-woven made of cellulosic fibres, is highly wettable when untreated, easy to functionalise and engineer, biodegradable, sterilisable, biocompatible and cheap. It has long being used for analysis in chromatography [15, 16]. Thin coatings of polymers and inorganics can easily be achieved through standard wet-end addition, surface sizing, coating and other surface treatments common to the industries. This suggests that paper is a natural material of choice for the production of disposable bioassay devices and bioactive surfaces. The concept of paper bioassays is to rely on paper for the transport of fluids samples, biomolecule detection/reaction, and the communication of the event in a single step process. Whitesides pioneered paper microfluidic systems by developing hydrophobic barriers on paper [17-19]. The pattern accuracy and manufacturing efficiency of microfluidic papers can be improved with plasma treatments or inkjet printing [20]. However, the biotechnology industry has a limited understanding of the effect of paper structure and chemical composition on the biomolecule functionality.

Eenzymes are model biomolecules of choice for a fundamental study because of their stereospecificity, wide range and availability. Understanding the effect of adsorption on the secondary and tertiary structure of enzyme, its functionality, stability and the selectivity of substrate recognition is critical for engineering biocatalysis. This knowledge can later be applied to other biomolecules. Few scientific studies have described the immobilization of enzymes on paper. Among those are paper strips indicators for sea food freshness using two enzymes (xantine oxidase and nucleoside phosorylase) [21], and bioactive paper to monitor alcohol content in the breath using alkaline oxidase (AOD) [22]. Chen et al. investigated by AFM the effect of surface wettability on enzyme loading (candida antartica lipase B CALB) [23]. However, there is little fundamental understanding on the effect of surface property on biofunctionality.

The first requirement of a bioactive paper is to retain the functionality, reaction rate and selectivity of its immobilized biomolecules. The second is to provide good biomolecule stability. The objective of this study is to measure the potential of paper as stable support for the immobilization of enzymes required for catalytic and diagnostic applications. Alkaline phosphatase (ALP) and horseradish peroxidase (HRP) were the enzymes selected for their stability and their wide range of applications.

In the first part of the study, the reaction kinetics of ALP adsorbed on paper is measured and modelled. In the second, a complete enzymatic pattern of HRP is printed on paper using thermal ink jet printing. It is the objective of the study to engineer fully bioactive and stable enzymatic papers.

Methodology

Materials

Horseradish peroxidase (HRP) and alkaline phosphatase (ALP) enzymes were purchased from Aldrich and used as received. HRP was used to formulate bio-ink and ALP to be immobilized on paper to study reaction kinetics of enzymatic bioactive papers. HRP was dissolved in 100mM sodium-phosphate buffer solution (pH 6.0) to a concentration of 1.0mg/mL to print on paper. A liquid substrate system, 3,3'-Diaminobenzidine (DAB), from Aldrich, was used to identify the enzymatic activity of the printed HRP. ALP was dissolved to a concentration of 0.1mg/mL in 1.0M diethanolamine buffer with 0.50mM magnesium chloride and 5M HCl to maintain pH at 9.7. Water (Millipore, 18MΩ) was used for making all dilutions. The 5-bromo-4chloro-3indoxyl phosphate and 3,3'-(3,3'-dimethoxy-4,4'-biphenylene)-bis-(2-p-nitrophenyl-5-phenyl-2H-tetrazolium chloride) liquid substrate system (BCIP/NBT), purchased from Aldrich, was used to quantify the enzymatic activity of ALP on paper. A filter paper (Whatman #4) was used as paper substrate to print on and immobilize biomolecules. Alkyl ketene dimer (Wax 88 konz, BASF) and alkenyl ketene dimer (Precis 900, Hercules Australia Pty Ltd) were used as cellulose hydrophobization agents. Analytical grade n-heptane (Aldrich) was selected as the solvent to make solutions of both dimers.

Enzyme on Paper

Enzyme aqueous solutions were homogeneously applied onto paper following TAPPI standard (T 205 sp-95). Paper samples (d = 16 cm) were immersed into the ALP enzymatic solution contained in a large Petri-dish (Figure 1). The well-soaked paper samples were then blotted and left dried in a dark chamber at standard condition: 23°C and 50% relative humidity (RH), for 24 hours. Experimental details, surface profile analysis and histogram distribution of gray values are described elsewhere [24].

Reactivity of ALP Enzymatic Papers

The liquid substrate was applied onto the enzymatic papers at t=0. This was done as follow: small droplets of fresh liquid substrates were applied onto the aged enzymatic papers using a 1.0 mL syringe equipped with a stainless steel flat-tipped needle (0.21 mm outer diameter). The enzyme-substrate (E-S) reaction was allowed to proceed at standard condition, 23ºC and 50% RH; this insures complete enzymatic reaction [24]. The colour change due to E-S reaction was captured using a standard imaging system, which is discussed in later section. From the colour intensity of E-S reaction on paper, the relative activity of the enzymatic paper was measured. Each measurement reported results from the average of 6 full replicates.

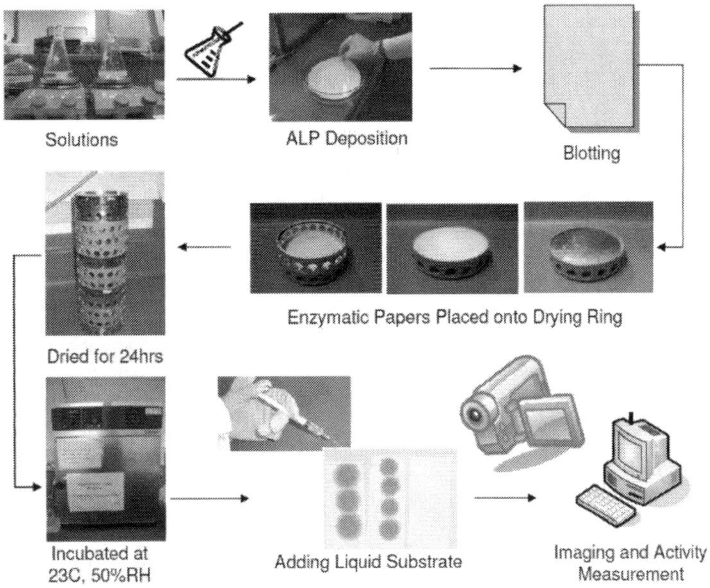

Figure 1: Experimental system to prepare ALP enzymatic papers.

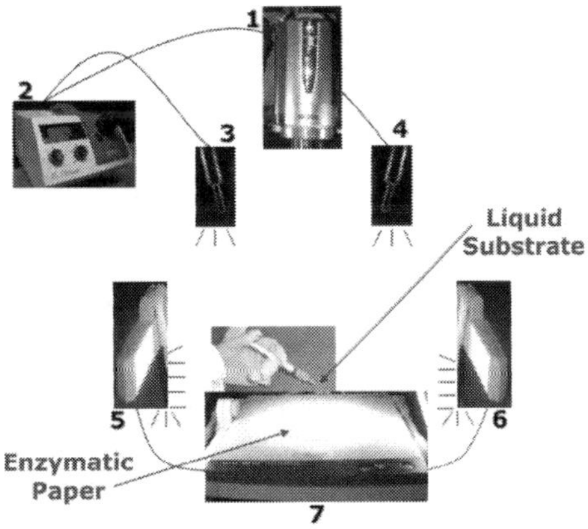

Figure 2: Experimental setup to measure enzymatic reaction kinetics on paper. 1) video camera, 2) light box, (3)-(6) lights, 7) platform for the ALP enzymatic papers.

Image Capturing and Activity Measurement

The reaction kinetics images were captured using a standard video camera (JVC Camcoder Everio GZ-MG530) (Figure 2). Two sets of lighting systems (SCHOTT

KL1500LCD, Germany and ELMO Visual Presenter, EV-2500AF PAL, Japan) were used for better resolution. The video clips were transferred to a computer and converted using Prism Video Converter v 1.27 and VirtualDub-1.8.8 software.

The converted images were analyzed using ImageJ software (ImageJ 1.41o). ImageJ calculates the gray values of RGB images. For any selected area, the ImageJ software calculates the weighted average gray value within the selection, which can be related to the activity of enzymatic paper.

Bio Printing

A basic Canon inkjet printer (Pixma ip4500) and ink cartridges (CLI, Y-M-C-BK, PGBK model) were reconstructed to print the hydrophobization solution and the HRP-enzyme solution on paper. This bubble jet printer offers a resolution of 9600×2400 dpi and is controlled by a personal computer which supplies page-data using MS Office 2003 software. A paper-based microfluidic pattern was created using the method describes elsewhere [25, 26]. HRP was printed inside the fluidic channels of the pattern. Details of printer modification, bio ink formulation and printing technique are discussed elsewhere [5].

Theory

Enzymatic Reactivity of ALP

Indoxy tetrazolium method [27] was used to investigate the kinetics of non-specific ALP. Combination of BCIP and NBT were used as the substrate. Together they yield an intense, insoluble dark-purple precipitate when reacted with ALP in alkaline condition. The complete reaction proceeds at a steady rate, allowing accurate control of the relative sensitivity and control of the development of the reaction [28].

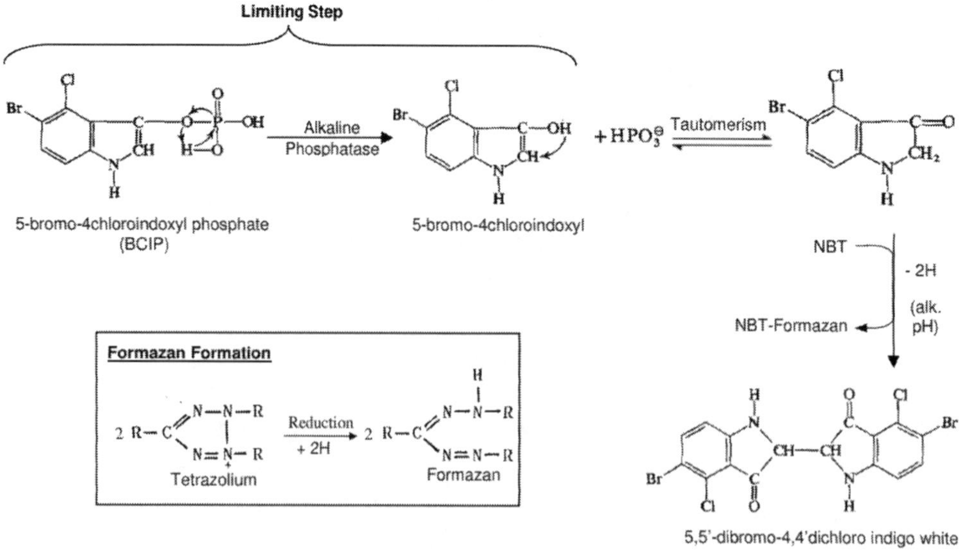

Figure 3: Kinetic scheme of the indoxy tetrazolium method for the histochemical demonstration of non-specific ALP. [27, 29]

Mechanism

BCIP is hydrolyzed by the ALP to form an intermediate, which at alkaline condition dimerises to produce an indigo dye: 5,5'-dibromo-4,4' dichloro indigo white. At alkaline condition the dimerisation reaction involves the loss of two hydrogen atoms, and these are used to reduce a tetrazolium salt 'NBT' to 'NBT- diformazan' at the enzyme site (Figure 3) [27, 30]. The diformazan is insoluble in alcohols and can be dehydrated, mounted in a synthetic medium and thus a permanent specimen obtained [27]. Step 2 represents tautomerism of BCIP intermediate, which is a fast reaction ($k \approx$ 3-5$\times 10^9$ $M^{-1}s^{-1}$) [31]. The redox reaction of NBT and BCIP intermediate to form NBT-formazan and indigo white dye is also a rapid reaction [30, 32]. Rate constant for NBT reduction to formazan varies from $10^6 - 10^7$ $M^{-1}s^{-1}$ [32]. The first step of the reaction scheme, which is hydrolysis of BCIP by ALP, is the rate limiting step of the reaction scheme (k = 30-118 s^{-1}) [33-35].

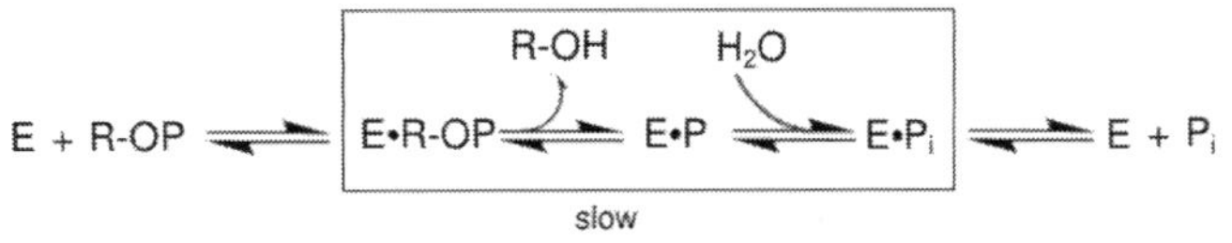

Figure 4: Mechanism of the ALP reaction [29, 33, 35, 36]

Figure 4 shows the hydrolysis mechanism of phosphate monoesters by ALP, where E is the free enzyme, R-OP is the substrate, E•R-OP is the activated enzyme-substrate complex, E•P is the phophorylated enzyme, and P_i is inorganic phosphate [34]. The first step represents ligand bonding [36]. The second step represents the catalytic reaction. The covalent E-P intermediate is subsequently hydrolyzed to give a non-covalent enzyme-phosphate complex (E•Pi). Wilson et al. [29] and Hull et al. [37], demonstrated that that the hydrolysis of the non-covalent E-Pi is the rate limiting under alkaline condition. Detailed discussion of ALP reaction mechanism can be found in [33-35].

Kinetic Model

In simplest form the E-S reaction can be represented by Eq.1:

$$\begin{array}{lll} & E+S \xrightarrow{k} & P+E \quad (1) \\ t=0 & a & 0 \\ t=t & a\text{-}x & x \end{array}$$

For the zero, first and second order reaction, kinematic models can be derived expressed from the above equation (Eq. 1), which are expressed in Eq. 2, 3 and 4, respectively [38]:

$$x = k\,t \quad (2)$$
$$\ln (a\text{-}x) = -\,k\,t + c \quad (3)$$
$$1/(a\text{-}x) = k\,t + c \quad (4)$$

Result and Discussion

Reaction Kinetics of ALP Enzymatic Papers

Liquid substrate was applied onto the ALP enzymatic paper samples and allowed the E-S reaction proceed to completion. The activities of enzymatic papers were calculated from the weighted mean grey value of the captured images. Figure 5 and 6 show the colorimetric evolution of enzymatic reactivity of ALP papers as a function of time. Pictures of product colour formation on ALP enzymatic papers are shown in Figure 5.

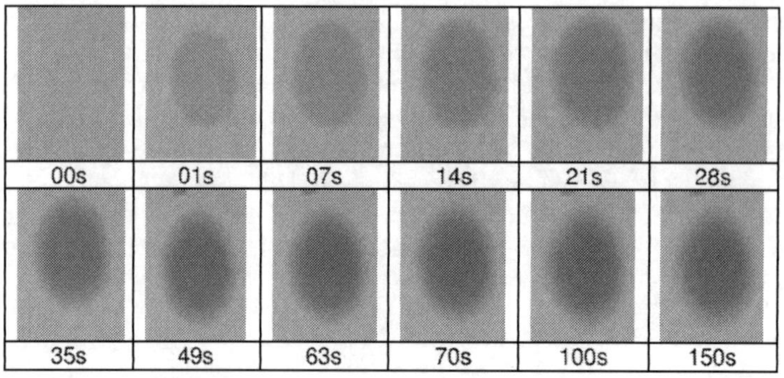

. Figure 5: Product formation on ALP enzymatic paper at different time. The blue purple colour reveals the enzyme (ALP) - substrates (BCIP/NBT) reaction.

Figure 6 illustrates the reaction kinetic model of ALP enzymatic paper. The product colour intensity (x) was found increasing up to first 40 sec. After that the colour changing became slow and gradually approached to saturation (Figure 6 a, b). Figure 6(c) shows that E-S reaction on ALP enzymatic paper follows the first order reaction kinetics. From the reaction mechanism of ALP with BCIP/NBT we found that hydrolysis of BCIP by ALP is the rate limiting step of ALP-BCIP/NBT reaction. The reaction rate constant (*k*) for ALP enzymatic paper was found 150 hr^{-1}. Scientists reported that the limiting step rate constant for ALP in buffer is 30-118 s^{-1} ($\approx$ 108 - 425 $\times 10^3$ hr^{-1}) [33-35]. Therefore, the ALP enzymatic activity was found 3 orders of magnitude faster in solution than that on paper.

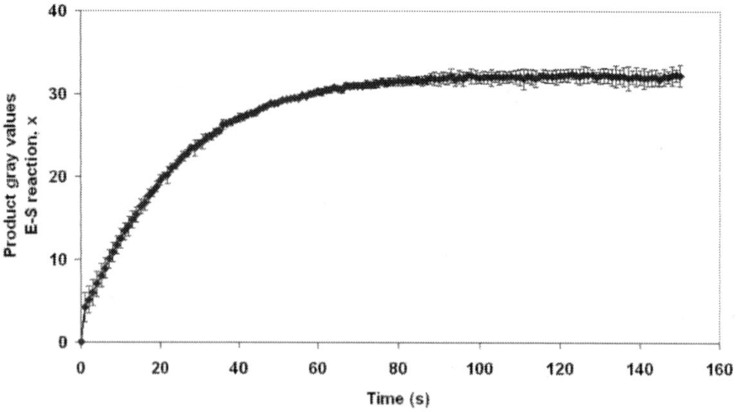

(a)

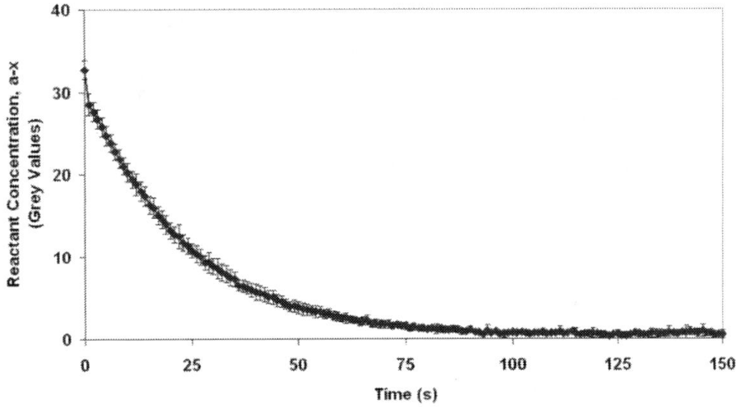

(b)

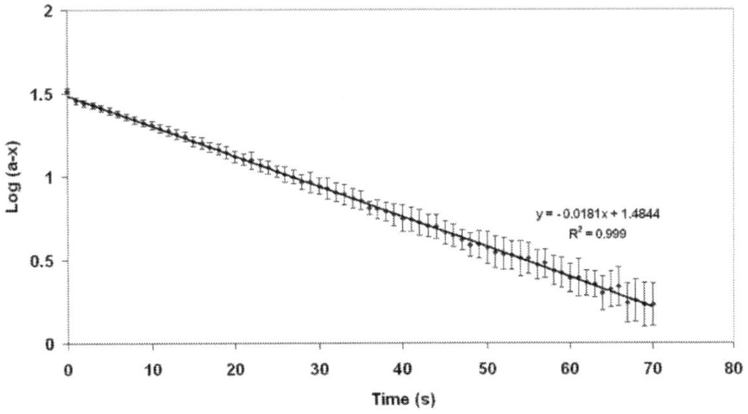

(c)

Figure6: Reaction kinetics of ALP enzymatic papers. (a), (b) are product and reactant concentrations as function of time in RGB gray value, respectively; (c) reactant concentration as a function of time in semi-log scale. The above model follows first order reaction kinetics.

Bio-Printing

Ink jet printing was used to fabricate a complete paper-based bioassays and diagnostic devices. Ink jet printing was first used to print hydrophobic patterns on paper, forming a series of four-channel hydrophilic systems [25] (Figure 7a). HRP enzyme was printed inside the hydrophilic channels (Figure 7a). A drop of DAB, the liquid enzyme substrate, was deposited in the collector at the centre of the hydrophilic pattern into the device using a micropipette. As the DAB solution penetrates along all channels, the brown colour was formed, revealing the exact position and the activity of the printed HRP. Figure 7b represents the photo of the paper based bio-diagnostics achieved.

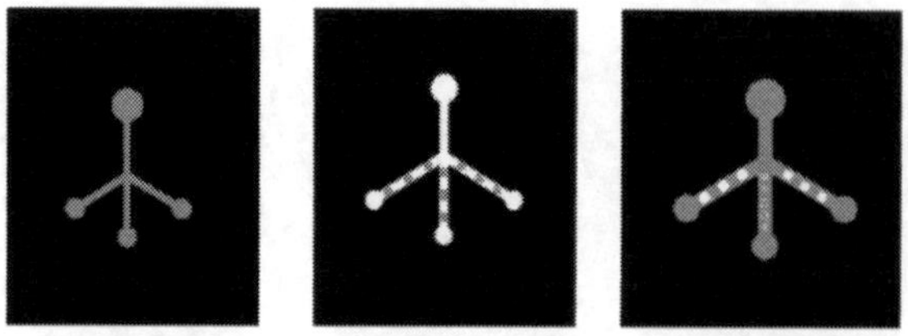

(a)

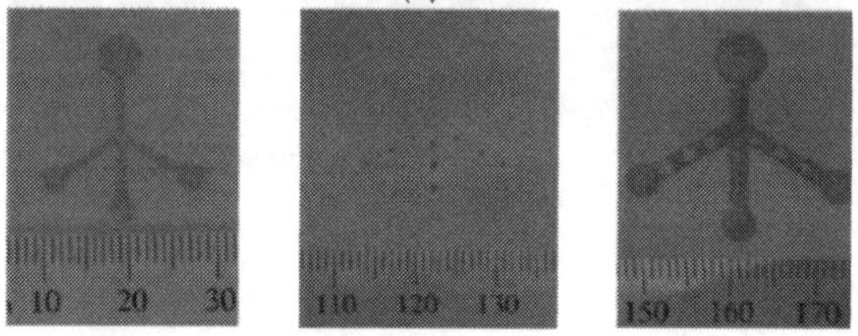

(b)

Figure 7: Paper based mircofluidic devices. (a) Microfluidic paper with three different designs of biomolecules patterns (magenta shades represent intended HRP printing patterns and black-white shades represent hydrophobic-hydrophilic contrast); (b) three different HRP printed patterns. The scales are in mm.

Figure 7 shows that enzyme sustained the printing thermal and shear stress and remained bioactive. This demonstrates that bioactive paper has exceptional potential for low cost, high flexibility diagnostic and industrial application.

Reference

[1] P. Calvert, Inkjet Printing for Materials and Devices, Chemistry of Materials 13 (2001) 3299-3305.

[2] P. Molnar, M. Hirsch-Kuchma, J.W. Rumsey, K. Wilson, J.J. Hickman, Biosurface Engineering, in: J.G. Webster (Ed.), Encyclopedia of Medical Devices and Instrumentation, John Wiley & Sons, 2006.

[3] D.R. Nisbet, S. Pattanawong, N.E. Ritchie, W. Shen, D.I. Finkelstein, M.K. Horne, J.S. Forsythe, Interaction of Embryonic Cortical Neurons on Nanofibrous Scaffolds for Neural Tissue Engineering, Journal of Neural Engineering 4 (2007) 35-41.

[4] A. Offenhausser, S. Bocker-Meffert, T. Decker, R. Helpenstein, P. Gasteier, J. Groll, M. Moller, A. Reska, S. Schafer, P. Schultea, A. Vogt-Eisele, Microcontact Printing of Proteins for Neuronal Cell Guidance, Soft Matter 3 (2007) 290-298.

[5] M.S. Khan, D. Fon, X. Li, J. Tian, J. Forsythe, G. Garnier, W. Shen, Biosurface Engineering Through Ink Jet Printing, Colloids and Surfaces B: Biointerfaces (In Press).

[6] L.R. Allain, M. Askari, D.L. Stokes, T. Vo-Dinh, Microarray Sampling-Platform Fabrication Using Bubble-Jet Technology for a Biochip System, Fresenius Journal of Analytical Chemistry 371 (2001) 146-150.

[7] S.M.Z. Hossain, R.E. Luckham, A.M. Smith, J.M. Lebert, L.M. Davies, R.H. Pelton, C.D.M. Filipe, J.D. Brennan, Development of a Bioactive Paper Sensor for Detection of Neurotoxins Using Piezoelectric Inkjet Printing of Sol-Gel-Derived Bioinks, Analytical Chemistry 81 (2009) 5474-5483.

[8] S. Ilkhanizadeh, A.I. Teixeira, O. Hermanson, Inkjet Printing of Macromolecules on Hydrogels to Steer Neural Stem Cell Differentiation, Biomaterials 28 (2007) 3936-3943.

[9] T. Jabrane, J. Jeaidi, M. Dube, P.J. Mangin, Gravure Printing of Enzymes and Phages, in: Iarigai (Ed.), Advances in Printing and Media Technology, 2008, pp. 279-288.

[10] M. Nakamura, A. Kobayashi, F. Takagi, A. Watanabe, Y. Hiruma, K. Ohuchi, Y. Iwasaki, M. Horie, I. Morita, S. Takatani, Biocompatible Inkjet Printing Technique for Designed Seeding of Individual Living Cells, Tissue Engineering 11 (2005) 1658-1666.

[11] J.D. Newman, A.P.F. Turner, Ink-jet Printing For The Fabrication of Amperometric Glucose Biosensors, Biosensors and Bioelectronics 20 (2005) 2019-2026.
[12] S.D. Risio, N. Yan, Piezoelectric Ink-Jet Printing of Horseradish Peroxidase: Effect of Ink Viscosity Modifiers on Activity, Molecualr Rapid Communications 28 (2007) 1934-1940.
[13] E.A. Roth, T. Xu, M. Das, C. Gregory, J.J. Hickman, T. Boland, Inkjet Printing for High-Throughput Cell Patterning, Biomaterials 25 (2004) 3707-3715.
[14] W. Zhao, A.v.d. Berg, Lab on Paper, Lab on a Chip 8 (2008) 1988-1991.
[15] K. Macek, Pharmaceutical Application of Thin-Layer and Paper Chromatography, Elsevier, 1972.
[16] L. Lepri, A. Cincinelli, TLC Sorbents, in: J. Cazes (Ed.), Enclyclopedia of Chromatorgrahpy, CRC Press, 2004, pp. 1-5.
[17] D.A. Bruzewicz, M. Reches, G.M. Whitesides, Low-Cost Printing of Poly(dimethylsiloxane) Barriers To Define Microchannels in Paper, Analytical Chemistry 80 (2008) 3387-3392.
[18] A.W. Martinez, S.T. Phillips, M.J. Butte, G.M. Whitesides, Patterned paper as a platform for inexpensive, low-volume, portable bioassays, Angewandte Chemie International Edition 46 (2007) 1318-1320.
[19] A.W. Martinez, S.T. Phillips, E. Carrithe, S.W. Thomas III, H. Sindi, G.M. Whitesides, Simple Telemedicine for Developing Regions: Camera Phones and Paper-Based Microfluidic Devices for Real-Time, Off-Site Diagnosis, Analytical Chemistry 80 (2008) 3699-3707.
[20] X. Li, J. Tian, T. Nguyen, W. Shen, Paper-Based Microfluidic Devices by Plasma Treatment, Analytical Chemistry 80 (2008) 9131-9134.
[21] G. Masao, Preparation of Enzyme-Immobilized Filter Paper for Determination of Freshness of Fish Meat, Jpn Kokai Tokkyo Koho, Japan, 1989.
[22] Y. Akahori, H. Yamazaki, G. Nishio, H. Matsunaga, K. Mitsubayashi, An alcohol gas - sensor using an enzyme immobilized paper, Chemical Sensors 20 (2004) 468-469.
[23] B. Chen, N. Pernodet, M.H. Rafailovich, A. Bakhtina, R.A. Gross, Protein Immobilization on Epoxy-Activated Thin Polymer Films: Effect of Surface Wettability and Enzyme Loading, Langmuir 24 (2008) 13457-13464.
[24] M.S. Khan, L. Xu, W. Shen, G. Garnier, Thermal Stability of Bioactive Enzymatic Papers, Colloids and Surfaces B: Biointerfaces (In Press).
[25] X. Li, J. Tian, M. Khan, G. Garnier, W. Shen, Low Cost Paper-Based Microfluidic Devices by Ink Jet Printing of Hydrophilic Hydrophobic Contrast, (2009, *In preparation*).
[26] W. Shen, T. Junfei, X. Li, M. Khan, G. Garnier, Method of Fabricating Paper-Based Microfluidic Systems by Printing, Austrailan Provisional Patent, 2008905776, Novermber 7 (2008), Austraila, 2008.
[27] J. McGadey, A Tetrazolium Method for Non-Specific Alkaline Phosphatase, Histochemie 23 (1970) 180-184.
[28] NBT/BCIP Substrates, in: T. Scientific (Ed.), Pierce Protein Research Products, Rockford, USA, 2009.
[29] I.B. Wilson, J. Dayan, The Free Energy of Hydrolysis of Phosphoryl-Phosphatase, Biochemistry 4 (1965) 645-649.
[30] F.P. Altman, Tetrazolium Salts and Formazans, Progress in Histochemistry and Cytochemistry 9 (1976).
[31] G. Dodin, O. Bensaude, J.-E. Dubois, Tautomerism of Formycin. Mechanism of Interconversion, Journal of The American Chemical Society 102 (1979) 3879-3899.
[32] O. Abugo, J.M. Rifkind, Oxidation of Hemoglobin and the Enhancement Produced by Nitroblue Tetrazolium, The Journal of Biological Chemistry 269 (1994) 24845-24853.
[33] J.E. Coleman, Structure and Mechanism of Alkaline Phosphatase, Annual Review of Biophysics and Biomolecular Structure 21 (1992) 441-483.
[34] D.B. Craig, E.A. Arriaga, J.C.Y. Wong, H. Lu, N.J. Dovichi, Studies on Single Alkaline Phosphatase Molecules: Reaction Rate and Activation Energy of a Reaction Catalyzed by a Single Molecule and the Effect of Thermal DenaturationThe Death of an Enzyme, Journal of The American Chemical Society 118 (1996) 5245-5253.
[35] K.M. Holtz, E.R. Kantrowitz, The Mechanism of the Alkaline Phosphatase Reaction: Insights from NMR, Crystallography and Site-Specific Mutagenesis, FEBS Letters 462 (1999) 7-11.
[36] N.C. Price, L. Stevens, Fundamentals of Enzymology, Oxford University Press, New York, 1984.
[37] W.E. Hull, S.E. Halford, H. Gutfreund, B.D. Sykes, ^{31}P Nuclear Magnetic Resonance Study of Alkaline Phosphatase: The Role of Inorganic Phosphate in Limiting the Enzyme Turnover Rate at Alkaline pH, Biochemistry 15 (1976) 1547-1561.
[38] B.S. Bahl, G.D. Tuli, A. Bahl, Essentials of Physical Chemistry, S. Chand & Company Ltd., New Delhi, 1999.

Rapid SERS Detection of Indirect Viral DNA Capture Using a Colloidal Gold Assay

Patrick A. Johnson, Chemical & Petroleum Engineering, University of Wyoming, Laramie, WY

Hao Zhang, Department of Chemical & Petroleum Engineering, University of Wyoming, Laramie, WY

Jing Neng, Chemical & Petroleum Engineering, University of Wyoming, Laramie, WY

Mark H. Harpster, Arthropod-Borne Animal Diseases Research Laboratory, USDA/ARS, Laramie, WY

William C. Wilson, Arthropod-Borne Animal Diseases Research Laboratory, USDA/ARS, Laramie, WY

Rapid and accurate detection is the key to controlling the spread of viral pathogens. Recent advances in the field of surface enhanced Raman scattering (SERS) spectroscopy have demonstrated the utility of this technology in diverse research disciplines ranging from surface chemistry and electrochemistry to forensics and the development of protein/nucleic acid sensor technologies. Our long-term goal is to develop a SERS based diagnostic tool for viremia that can be easily adapted for cost effective, portable and compact Raman spectroscopy in a point-of-care facility or field setting.

In this work, an indirect capture model assay using colloidal Au nanoparticles was demonstrated for SERS spectroscopy detection of DNA sequences and viral antigens. The DNA sequence targeted for the indirect capture was derived from the West Nile Virus (WNV) RNA genome and selected on the basis of exhibiting minimal secondary structure formation. Upon incubation with colloidal Au, hybridization complexes containing the WNV target sequence, a complementary capture oligonucleotide conjugated to a strong tethering group and a complementary reporter oligonucleotide conjugated to methylene blue, a Raman label, anchors the resultant ternary complex to Au nanoparticles. This anchoring positions the methylene blue within the required distance from the gold nanoparticles for SERS enhancement.

See more of this Session: Biosensor Devices I
See more of this Group/Topical: Topical 9: Sensors

Surface Enhanced Raman Scattering Based Biomolecular Sensing Techniques in Optofluidic Device

Yun Suk Huh[a], Adam J. Lowe[b], Aram J. Chung[a], Bernardo Cordovez[a], Aaron D. Strickland[c], Carl A. Batt[c], David Erickson[a,*]
[a] Sibley School of Mechanical and Aerospace Engineering, Cornell University, Ithaca, NY 14853, USA.
[b] Department of Microbiology, Cornell University, Ithaca, NY 14853, USA
[c] Department of Food Science, Cornell University, Ithaca, NY 14853, USA

I. Introduction

Recently, Surface-enhanced Raman scattering (SERS) has attracted great attention because of its high sensitivity, label-free detection, and low detection limit. Additionally, it provides a specific fingerprint spectrum of multiplex samples with narrow band width. Because of these advantages, a number of research groups have investigated the SERS based biomolecular detection using the conventional DNA-DNA hybridizations and protein-protein interactions. Here, we demonstrated new approaches to detecting 1) Single Nucleotide Polymorphisms (SNPs) based Ligase Detection Reaction (LDR) and 2) peptides of vasopressin based aptamer-protein interactions techniques.

II. Results

1) Surface Enhanced Raman Scattering based Ligase Detection Reaction

SNPs have recently become key indicators of genetic disease, allelic variation associated with cancer progression, and pharmcogenomics. There are several methodologies such as primer extension, oligonucleotide ligation, invasive cleavage or hybridization for SNP detection. Among them, LDR is more accurate than hybridization approaches, permits parallel analysis of several loci directly on genomic DNA, and requires very low amounts of detection oligonucleotides per genotyping reaction. Therefore, we present a novel SERS-LDR approach by combining LDR reaction with SERS where a Raman spectrum was diagnostic for positive ligation and SNP discrimination. Figure 1 shows a new SERS-LDR schematic representation. In this technique, the upstream primer contains a SERS active dye and the discriminating 3' base, while the downstream primer contains an amine to which a silver nanoparticle is attached. When the two primers are ligated together the dye is brought into close proximity to the nanoparticle and its Raman signature is detectable. Different Raman active fluorophores are placed on the discriminating primers based on their genotype, which produce a signature Raman profile as shown in Fig. 1.

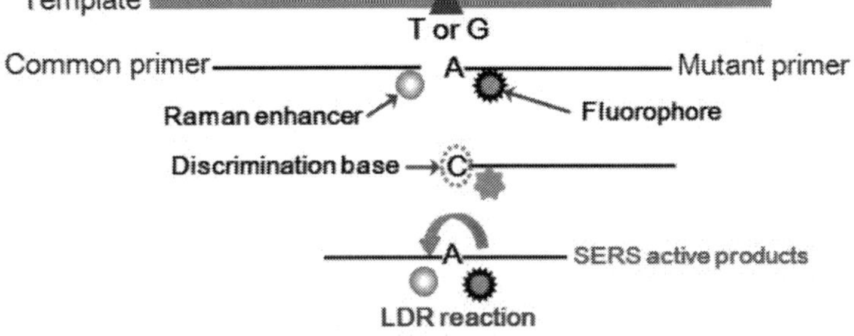

Figure 1. Overview of SERS Enhanced PCR/LDR detection reaction.

To obtain the faster and more sensitive SERS signals, we implement the SERS-LDR method in the optofluidic device that enables us to enhance SERS detection by concentrating the multiplex products from the bulk phase into a confined volume. For the on-chip assays, the SERS-LDR products are concentrated into the 10 μm wells by applying electric potential. Once concentrated, the well can be interrogated optically through the upper PDMS. To characterize the multiplex SNPs detection based SERS-LDR method, we examined the spectra collected on-chip at various ratios of wild type (TAMRA) to mutant type (fluorescein) templates (5:1, 3:1, 1:1, 1:3 and 1:5). The intensity of the peaks corresponds well with the ratios of the SNP templates. As expected, a near linear curve is generated. Using SERS-LDR products coupled with the electrokinetic device, we detected SERS signals for SNPs with a limit of detection on the order of 20 pM.

2) Aptamer based SERS of Vasopressin using Multilayer Nanotube Arrays

As another approach, we demonstrate the SERS detection of vasopressin (VP) based aptamer interaction technique. Vasopressin is a peptide hormone, which can serve to rapidly increase arterial pressure thereby stabilizing the patient. As such, VP can be considered both a biomarker of a critical injury state and a therapeutic agent. In this study, to rapidly and specifically monitor VP, we develop the aptamer based SERS detection approach in a SERS-active microfluidic device. Briefly, aptamers are artificial nucleic acid ligands that can bind to a target molecule such as amino acids, drugs, and proteins with high specificity and affinity in a similar manner of antibodies. However, aptamers have significant advantages in that they are synthetically derived, making production predictable unlike antibodies. Figure 2 shows the schematic of the aptamer-VP recognition binding reaction. Briefly, to immobilize the aptamer, the thiolated aptamers were added onto the SERS-active silver substrate in PBS buffer solution. And then, the FITC-labelled VP in PBS buffer solution was introduced into the sample inlet port of the SERS device. To characterize the reaction specificity, the fluorescence intensity of each spot was analyzed with the aptamer-VP reaction procedures. From the fluorescent image results, it can be clearly seen that the aptamer has its specific recognition function toward VP.

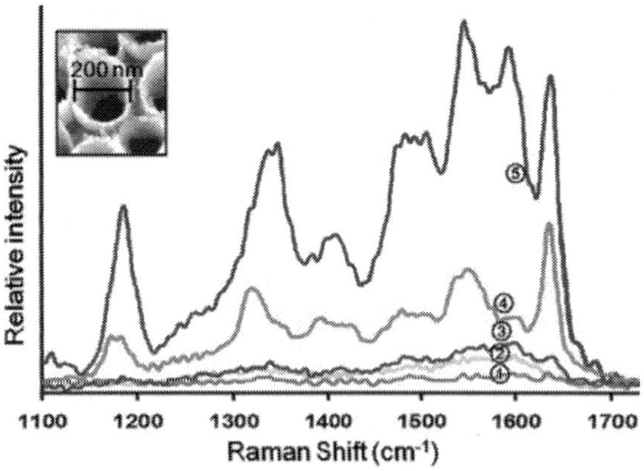

Figure 2. SERS spectra of aptamer-VP reaction corresponding to the first 3 cases described shown in image (b) and (4) for 200 nm Au nanotube arrays and (5) Au/Ag/Au nanotube arrays.

To improve the reproducibility, we examined the SERS spectra of VP based aptamers in a microfluidic device pattered 200 nm metal nano-tube array. For the fabrication of SERS-active chip, Au/Ag/Au (5 nm/190 nm/5 nm) was deposited onto the walls of the AAO membrane pores at an incident angle 45. After evaporation, dissolving the AAO in 3 M NaOH solution removes the template to obtain an array of nanotubes connected by a thin backing of the same metal, followed by changing the dissolving solution to the 3-mercaptopropyl-trimethoxysilane (MPTMS) solution. The MPTMS layer served as an organic adhesion layer aiding with the bonding between nano-tube array and PDMS

substrate. For the bonding of the nano-tube array patterned PDMS and the PDMS channels patterned by soft-lithography, the surfaces of both layers were activated in oxygen plasma.
To verify the selective binding assay of aptamer, we conducted SERS detection experiments according to the reaction procedures based apatmer (Fig. 2a). Figure 2b shows the SERS spectra collected from on-chip nano-tube arrays region for (1) background control of nano-tube array substrate, (2) aptamers immobilized onto the nano-tube array, (3) FITC-labeled bovine serum albumin (BSA) (negative control), and (4) FITC-labeld VP (positive control). As can be seen in Fig. 2b(3), our results show that small peak around 1600 cm^{-1} was observed from the negative control sample, but there are almost no detectable SERS signals beyond 1600 cm^{-1}. For the positive control sample, the SERS spectra of VP are generated from the SERS active molecules located in the region of the nano-tube structures. Figure 2b(4) shows the correct spectroscopic fingerprints corresponding to FITC-labeled dye suggesting positive detection. To verify the ability for quantitative analysis, we conducted a series of experiments at different VP concentrations. The intensity of the peaks at 1823 cm^{-1} corresponds well with the increase of VP concentrations. Using the aptamer as VP recognition agent, we successfully detected SERS signals and demonstrated the ability to quantify the solution concentration based on the linearity of SERS intensity.

DNA Biosensors for the Detection of Campylobacter Jejuni Nucleic Acids Based On SPR and Laser Diffraction Platforms

Tony Gnanaprakasa, Materials Research and Education Center, Auburn University, Auburn, AL

Valber A. Pedrosa, Materials Research and Education Center, Auburn University, Auburn, AL

Eric V. Olsen, Clinical Research Laboratory, Keesler AFB, Biloxi, MS

Omar A. Oyarzabal, Department of Poultry Science, Auburn University, Auburn, AL

Aleksandr L. Simonian, Materials Research and Education Center, Auburn University, Auburn, AL

DNA-biosensors allow rapid, real-time monitoring of hybridization with the target nucleic acids. These biosensors are gaining more attention over traditional diagnostic methods for their fast and responsive operation, and their cost effective design. Specificity of the system relies on oligonucleotide probes covalently immobilized on the sensing surface. We report the development of a biosensor based on direct covalent coupling of thiol and biotin labeled single-stranded DNA (ssDNA) on a Surface Plasmon Resonance (SPREETA™) and a laser optics (dotLab™) transduction platform to detect genomic DNA for Campylobacter jejuni. The laser optics platform (dotLab™) has been used for the first time to detect ssDNA target. The sensors used in this system consist of a polystyrene surface with neutravidin coated spots over which the biotin-labeled DNA is immobilized. The detection limit of both the sensor platforms has been analyzed and it was found that the dotLab™ system shows better detection limit up to 1 nM than the SPR system which has a detection limit of 10 nM. The reusability of the sensors has also been analyzed, and we found that there was negligible change in sensitivity after sensors regeneration (~9.7 x E-7 ΔRI) with minimal damage to the immobilized probes. Non-specific binding was minimized using a spacer thiol to the thiolated probes and BSA in the laser optics system, to cover the non-avidinated polystyrene region. The spacer thiol, not only removes unbound ssDNA, but also creates a self-assembled monolayer thereby minimizing non-specificity. These findings highlight the potential of DNA biosensors in developing a sensitive, regenerative and robust platform for food safety and clinical investigations.

See more of this Session: Biosensor Devices I
See more of this Group/Topical: Topical 9: Sensors

DNA Microarrays Formed by Microquill Guided Attachment to Chemically Synthesized Oligonucleotide Surfaces

Zhou Xu, Department of Chemical and Biomolecular Engineering, Vanderbilt University, Nashville, TN

Paul E. Laibinis, Department of Chemical and Biomolecular Engineering, Vanderbilt University, Nashville, TN

DNA microarrays have become an important tool for studies of gene expression, single nucleotide polymorphisms, and comparative genomics. These immobilized patterns of DNA strands on support have been prepared by a variety of methods, differing in their ease of preparation, their specificities, and sensitivities. A frequently employed in-house approach is the spotting of gene-specific PCR products onto a poly-L-lysine coated microscopic slides. This method electrostatically fixes the DNA strands to the surface, albeit by multiple points along the DNA sequence. End-immobilized DNA sequences have instead been fabricated by light-directed combinatorial chemical synthesis and by inkjet printing technology, both requiring highly specialized equipment for their construction. We have developed an alternative method for preparing DNA microarrays, taking advantages of both oligonucleotide synthesis chemistry and conventional spotting methods for their in-house preparation. With the former method, a covalently bound single-stranded oligonucleotide sequence is formed one base unit at a time, by a fully selectable and programmable process. This sequence is designed to be specific toward a complementary sequence that is attached to an end region of oligos that are patterned onto the surface using a microprinting technique. The talk will describe key features of this approach: the synthesis of single stranded DNA sequences on glass substrates; the specificity of these surface DNAs toward the binding to complementary DNA and their rejection of non-complementary sequences, their use in generating microarray patterns that exhibit selectivity in DNA binding, and the quantitation of binding preferences by fluorescence methods. Relevant timescales of each steps and process integration will be discussed.

See more of this Session: Biosensor Devices I
See more of this Group/Topical: Topical 9: Sensors

Characterization of Antibody Immobilization Methods Using the QCM-D

Ashish S. Yeri, Department of Chemical and Petroleum Engineering, University of Pittsburgh, Pittsburgh, PA

Lizeng Gao, Department of Chemical and Petroleum Engineering, University of Pittsburgh, Pittsburgh, PA

Di Gao, Department of Chemical and Petroleum Engineering, University of Pittsburgh, Pittsburgh, PA

Biosensors sensors which incorporate biomolecules such as proteins, cells or DNA to detect the target molecule (analyte) find applications in numerous fields such as medical diagnostics, environmental applications and food analysis. The high specificity, sensitivity and rapid detection times have warranted great interest in the field of biosensors. Biosensors which utilize proteins such as antibodies or enzymes to sense the target molecules are usually immobilized onto substrates. But, the reliability of antibody based biosensors has been a major issue as immobilization of the antibodies on substrates tends to make them inactive, as a result of which the antibody loses its 'biosensing' ability. The reasons for inactivity could be due to the denaturation of the antibodies on the surface or because of the unfavorable orientation of the surface antibodies wherein the antibody is unable to bind to the antigen.

In this work, we examine some of the most commonly used techniques to immobilize antibodies and provide a quantitative means to evaluate the different immobilization chemistries. We have studied the immobilization of Mouse IgG antibody via three different immobilization schemes and its interaction with Antimouse IgG antibody. Mouse IgG Antibodies were immobilized by covalent attachment to the substrate (randomly oriented), attachment via Protein A which specifically binds to the Fc region of the antibody (oriented) and finally, attachment by Biotinlyated antibody to a Streptavidin covered substrate (oriented). Equilibrium adsorption studies were conducted for each of the immobilization chemistries using the quartz crystal microbalance (QCM), a surface acoustic wave based sensor. The QCM can measure extremely low masses (ng/cm2) with very high precision by accurately measuring the decrease in the resonant frequency of the quartz crystal when mass is deposited on it. The frequency of resonance of the quartz crystal decreases linearly with the mass adsorbed (Saurberey's equation) and so, a very accurate and sensitive measurement of mass adsorbed can be made. The equilibrium adsorption isotherm between the Antibody and Antigen follows the conventional Langmuir Adsorption model, from which we can obtain useful information about the association equilibrium constant (Ka) and the surface concentration (Ab0) of active antibody. From the results of our work, we can conclusively say that the surface concentration of active antibodies is higher for the oriented immobilization of antibodies as compared to the randomly oriented antibody, which makes it more suitable for application in biosensors.

See more of this Session: Biosensor Devices I
See more of this Group/Topical: Topical 9: Sensors

Integrated Focused and Conventional Transducer Configuration On a Single Piezoelectric Device for Simultaneous Biosensing and Biofouling Removal

Reetu Singh, Department of Chemical & Biomedical Engineering, University of South Florida, Tampa, FL

Subramanian K.R.S. Sankaranarayanan, School of Engineering and Applied Sciences, Harvard University, Cambridge, MA

Venkat R. Bhethanabotla, Department of Chemical and Biomedical Engineering, University of South Florida, Tampa, FL

Summary Transducers used in biosensing applications are plagued by biofouling, which refers to the binding of non-specific proteins to the device surface resulting in a compromise of the device sensitivity and selectivity. Acoustic streaming, resulting from high intensity sound waves, has the potential to address the issue of biofouling elimination in biosensors [1]. Focused interdigital transducers (IDTs) have the potential for acoustic energy focusing, thereby increasing the intensity of acoustic streaming [2]. We have identified that in a Langasite substrate, the (0, 22, 90) Euler direction is suited for biosensing via the propagation of predominantly shear horizontal waves and the (0, 22, 0) direction allows for biofouling removal via mixed mode propagation with prominent surface normal component. To design a better biosensor, we develop a finite element model of a surface acoustic wave device with a multidirectional mutually interacting IDT configuration: uniform IDTs (U-IDTs) along (0, 22, 90) direction and focused IDTs (F-IDTs) along the (0, 22, 0) direction. The enhancement in sensor performance was analyzed in terms of device sensitivity and acoustic streaming force. Our results indicate that the streaming force and the sensitivity for the device with the mutually interacting U-IDTs/F-IDTs are significantly higher when compared to uniform unidirectional IDTs. This work broadly applies to all transducers used for biological species sensing that suffer from fouling and non-specific binding of protein molecules to the device surface. Computational details A three dimensional finite element model of a SAW device based on Langasite substrate, with U-IDTs/F-IDTs was developed. The dimensions of the piezoelectric substrate were 1600μm width x 1600μm propagation length x 200μm depth. The device was modeled with two port delay line consisting of two sets of IDTs along each of the two delay paths: the input IDTs and the output IDTs. The fingers were defined with periodicity of 40 μm and aperture width of 200 μm along (0, 22, 90) direction. Along the (0, 22, 0) direction, aperture width of the fingers varies depending on their radial distance from device center The model was meshed with tetragonal solid elements with four degrees of freedom, three of them being the three translations and the fourth being the voltage. To optimize on the computation time while capturing the dynamics accurately, highest mesh densities were ensured near the device surface and the middle of the substrate. A total of 388, 893 nodes and 266, 762 elements were generated. An impulse response analysis was performed during 190 ns by applying an impulse voltage of 100 V at the input IDT and employing a time step of 0.95 ns to deduce the central frequency of the device in each direction. The central frequencies were computed as 68 MHz and 64 MHz along the (0, 22, 90) and (0, 22, 0) direction, respectively. Subsequently, an AC analysis was carried out, by applying a peak voltage of 2.5 V to the input IDTs in each of the two Euler directions and employing a time step of 1 ns, to investigate the nature of waves propagating in the two directions. The device sensitivity and acoustic streaming force were computed and compared with devices having uniform unidirectional IDTs in the respective directions. Results Our previous experimental studies have demonstrated acoustic streaming can be effectively used to achieve removal of non-specifically bound (NSB) proteins from the device surface (Fig. 1), thereby enhancing device performance. We present the first report on mutually interacting U-IDTs/F-IDTs on a Langasite substrate to enhance device performance in biosensing applications by increasing the intensity of acoustic streaming forces for efficient biofouling removal and amplify the displacement/surface acoustic energy for increased device sensitivity. Our simulation studies indicate the propagation of pure shear horizontal wave along the (0, 22, 90) direction, which allow for biosensing. Focused IDTs along the (0, 22, 0) direction lead to the focusing of waves (Fig. 2) thereby maximizing the device displacements near the focal point. Further, the use of focused IDTs along (0, 22, 0) direction allows for amplification of the prominent surface normal component in the propagating mixed mode wave. This amplification of the surface normal component leads to a 266% increase in the amplitude of streaming force (4.2*105 N/m2 for U-IDTs/F-IDTs design vs. 1.2*105 N/m2 for standard uniform IDTs). In addition, the constructive interference of propagating waves in the delay path leads to maximized energy entrapment and displacements on the device surface, thereby imparting 32.4% higher sensitivity to the device with U-IDTs/FIDTs (15.3 Hz-cm2/ng) compared to the standard unidirectional IDT design (11.56 Hz-cm2/ng). Our simulation results indicate that the use of mutually interacting U-IDTs/F-IDTs on a Langasite substrate leads to an efficient sensor with increased device sensitivity and intensity of acoustic streaming compared to a device with uniform unidirectional IDTs. Results will be presented at the AIChE 2009 meeting.

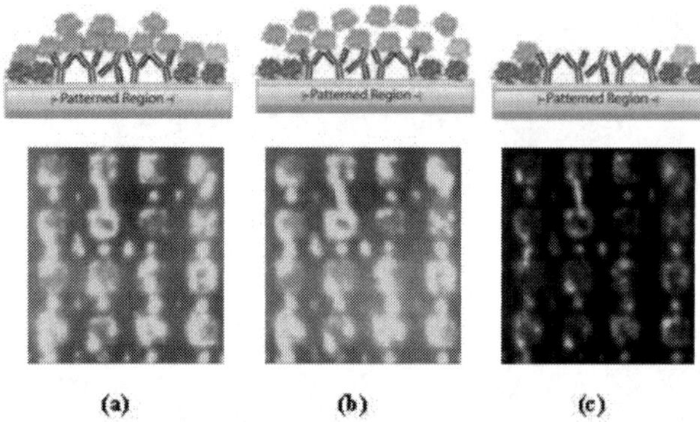

Fig. 1: Non-specifically-bound BSA fluorescent image (pseudo-colored) with illustration above of SAW device surface at end of experiment (RF power dose = 12,500 mW sec). The illustrations in (a), (b), and (c) depict the removal of non-specifically-bound proteins accomplished by acoustic-streaming phenomenon induced by the Rayleigh wave interaction with the fluid medium [3].

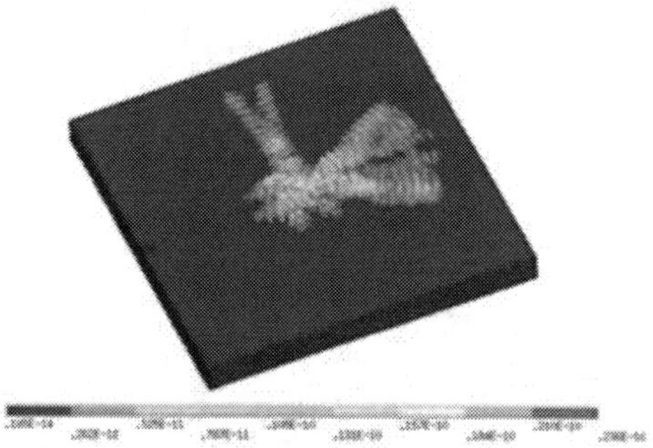

Fig. 2: Multidirectional IDT design for a SAW based on Langasite. Propagation direction (0, 22, 90) with conventional IDT has a pure shear horizontal mode which would be used for biosensing. Orthogonally located focused IDTs along (0, 22, 0) direction allow for amplification of surface normal component leading to significantly higher streaming forces. This allows for simultaneous removal of non-specifically bound protein as shown in Fig. 1.

References:

1. Subramanian K.R.S. Sankaranarayanan, Stefan Cular, Venkat R. Bhethanabotla and Babu Joseph. Physical review E, 77, 6 066308/1-19 (2008).

2. Subramanian K.R.S. Sankaranarayanan, and Venkat R. Bhethanabotla. Journal of Applied Physics, 103, 6, 064518/1-17 (2008).

3. S. Cular, D. W. Branch, V. R. Bhethanabotla, G. D. Meyer, and H. G. Craighead, IEEE Sensors Journal 8, 314 (2008).

See more of this Session: Biosensor Devices I
See more of this Group/Topical: Topical 9: Sensors

Resumption of Bioluminescent Gene Expression in Whole-Cell Bacterial Biosensors After Performing High Temperature Switch

Stephanie Fernandez, Chemical Engineering, Northeastern University, Boston, MA

Frank Marealle, Chemical Engineering, Northeastern University, Boston, MA

Jinyi Han, Chemical Engineering, Northeastern University, Boston, MA

Dennis M. Callahan Jr., Department of Chemical Engineering, Center for Advanced Microgravity Materials Processing (CAMMP), Northeastern University, Boston, MA

Kostia Bergman, Biology, Northeastern University, Boston, MA

Jacqueline Piret, Biology, Northeastern University, Boston, MA

Katherine S. Ziemer, Chemical Engineering, Northeastern University, Boston, MA

Albert Sacco Jr., Department of Chemical Engineering, Center for Advanced Microgravity Materials Processing (CAMMP), Northeastern University, Boston, MA

According to Frost and Sullivan, the global market for biosensors is projected to grow from $5.11 billion in 2006 to $10.62 billion in 2013 as biosensors have emerged as a leading technology in the application of analyte detection and analysis. Whole cell biosensors, which use living microbial organisms to detect changes in their environment, have been shown to exhibit the potential to both detect the presence of targeted chemical substances with high selectivity and determine their concentrations with high sensitivity (in qualitative and quantitative sensing properties).[1] The Pseudomonas putida strain TVA8 bacterial-based biosensor is used to sense parts-per-billion levels of chemicals or inducers (i.e. trichloroethylene, TCE) through bioluminescence. However, the technical challenge in detecting changes in inducer concentrations is to engineer an accurate quantitative link between the bioluminescence signal and the inducer concentration.

It has been shown previously that the bioluminescence signal can be "turned off" or "reset" utilizing temperature.[2] It would be beneficial to continuously "turn-off" the bioluminescence signal so that the signal from the biosensor can be "re-set" after receiving a measurable response of TCE concentration for calibrating the bioluminescence signal. Previously discovered was a method for "turning off" the bioluminescence in P. putida strain TVA8, through controlled heating to denature proteins involved in the bioluminescence mechanism. The bacteria were grown at 27 °C on TCE-saturated minimal media shaking at 220 RPM. At 8 hours, the temperature of the bacteria was increased to 37 °C using a water bath. The bioluminescence was recorded once per hour prior to the temperature change and once per minute after the change. After raising the temperature, the bioluminescence signal was found to decrease rapidly, with an average of 95.388 ± 1.59% of the maximum bioluminescence eliminated after 2 minutes. Therefore, it is possible to eliminate the bioluminescence signal using a high temperature (37 °C) control "switch", which is necessary for sensor calibration.

The goal of this study is to verify this high temperature control "switch" can be used to calibrate the bioluminescence signal. Additionally, resumption of the bioluminescence signal is tested after the high temperature (37 °C) control "switch" is performed for calibration purposes as well. The bacteria were grown at 27 °C on TCE-saturated minimal media for 7 hours. After the 7th hour, the high temperature (37 °C) control "switch" was performed for approximately 30 minutes causing the bioluminescence signal to "turn off" after 15 minutes. Afterwards, the temperature of the bacteria is decreased back to 27 °C and after 30 minutes the bioluminescence signal resumed. Resumption of the bioluminescence signal shows that the signal can be calibrated in order to read multiple events in succession.

[1] Applegate, B.M., et al. Applied and Environmental Microbiology, Vol. 64, No.12, 1998.

[2] Callahan, D., et al., "Inhibition of bioluminescent gene expression in whole-cell bacterial biosensors using a high temperature switch," AIChE National Conference, San Francisco, CA, 2006.

See more of this Session: Biosensor Devices II
See more of this Group/Topical: Topical 9: Sensors

Post-Script Fabrication of Bioseparation Membranes within Sealed, Completed Microfluidic Devices

Lucas D. McIntosh, Department of Chemical Engineering, Missouri University of Science and Technology, Rolla, MO

Chang-soo Kim, Electrical & Computer Engineering, Missouri University of Science and Technology, Rolla, MO

David B. Henthorn, Department of Chemical & Biological Engineering, Missouri University of Science and Technology, Rolla, MO

In recent years, there has been an effort to develop self-calibrating electrochemical and optical biosensors that require minimal human intervention. Separation membranes are structural elements that will feature prominently on many of these biosensors, and fabrication techniques which allow for the formation of membranes in sealed channels are especially interesting. Membranes polymerized in completed microchannels can form tight seals between the floor and ceiling, especially if these surfaces have been chemically modified. Additionally, if membranes are synthesized before the microchannels are sealed, the wafer bonding process used to seal the channel may require temperatures that are not amenable to the membrane materials, particularly if functionalization with biomolecules is used. Traditional photolithographic methods which employ photomasks suffer from several disadvantages brought about by the long working distance between mask and membrane precursor solution due to the thickness of the capping wafer. In this work, we detail the fabrication of membranes in completed, sealed microchannels through the use of a novel "post-script" procedure which uses on-chip photomasks.

See more of this Session: Biosensor Devices II
See more of this Group/Topical: Topical 9: Sensors

Microfluidic Synthesis of Silica Particles and Bioconjugation

Gerson R. Aguirre, Chemistry Department, City College and Graduate Center of the City University of New York, NY, NY

Alex Couzis, Chem. Eng., City College and Graduate Center of the City University of New York, NY, NY

Charles Maldarelli, Levich Institute, City College and Graduate Center of the City University of New York, NY, NY

Shahab Shojaei-Zadeh, Levich Institute, City College and Graduate Center of the City University of New York, NY, NY

Entrapment and retention of materials inside synthesized silica particles is an active research field due to its many industrial, biological and chemical activities including as molecular sieves due to their porosity. Many methods are used to create these particles, however, most require the removal of the templating agent by acid burning or calcining of the organic material which can be detrimental if a sensitive material needs to be preserved. What follows is a fabrication scheme by use of microfluidic channels that will allow particle synthesis and a narrow size distribution, allowing for retention of introduced material or molecules inside the silica particle that can be beneficial in biological applications.

We report on the formation of silica particles with entrapped fluorescent semiconductor CdSe/ZnS quantum dots (QDs) by a fast synthesis method within a microfluidic cell. The hydrophobic nature of the QD surface stabilizes their dispersion in the oil phase. The synthesis of TEOS (tetraethyl orthosilicate), in the presence of Octylamine, by a highly acidic solution within a few minutes gives micronsized particles. The distributed oil phase provides the vestibule for the QDs and as the silica particle is formed at the oil-water interface the QDs are trapped in the core. Appropriate extent of hydrolysis is obtained and gelation at the oil/water interface occurs by the driving force of condensation, possibly lead by the alkylamine. Confocal microscopy images confirm the quantitative capture of the QDs within the formed silica particle. Additionally, we have bioconjugated said Silica particles with antigens, and other proteins.

See more of this Session: Biosensor Devices II
See more of this Group/Topical: Topical 9: Sensors

Orthogonal SAW Device Based On Langasite for Simultaneous Biosensing and Biofouling Removal: A Fluid Structure Interaction Study

Reetu Singh, Department of Chemical & Biomedical Engineering, University of South Florida, Tampa, FL

Venkat R. Bhethanabotla, Department of Chemical and Biomedical Engineering, University of South Florida, Tampa, FL

Summary: Multi-directional transducers have the capability of achieving the dual objectives of biosensing and non-specifically bound protein removal for improved sensor performance. Understanding the acoustic wave propagation and the sensing mechanism in SAW biosensors involving fluid interactions with complex multi-directional transducers represents a significant challenge. Till date, attempts to address this have relied on simplified numerical and analytical models or perturbation theories which neglect the mechanical properties of the fluid and treat the leaky wave as a first-order perturbation on the non-leaky wave associated with surface-wave propagation1. We present the first report on a three dimensional fluid-structure interaction of SAW sensor utilizing a multi-directional transducer based on a Langasite substrate. Numerical solutions are obtained by sequentially solving the generalized Navier-Stokes equation and acousto-electric equations for the solid motion2. Precise knowledge of the induced streaming phenomenon due to the developing flow fields and wave propagation characteristics arising from the transient fluid-piezoelectric interaction is provided by the developed 3-D models. These are utilized to compute streaming forces for predicting mechanism of biofouling elimination in acoustic devices. The results are of tremendous significance for not only improving the device design, but also for understanding biosensing mechanisms in multi-directional acoustic wave devices as well as actuation mechanisms in potential microfluidic applications of these devices. Computational details: In this work, a novel biosensor based on a Langasite substrate is investigated to analyze its suitability for biosensing applications, using combined structural and fluid-solid interaction finite element models. The fingers were defined with periodicity of 40 µm along both the propagation directions. Quadrilateral coupled field solid elements were used to model the solid piezoelectric domain. Impulse and AC analysis was carried out using structural models to determine the device frequency response and wave propagation characteristics. Additionally, coupled field fluid-structure interaction (FSI) model was developed to study surface acoustic wave interaction with fluid loading. Specifically, finite element models of a SAW device based on a micron sized LGS (0, 22, 0) piezoelectric substrate in contact with a liquid loading were developed and solved to gain insights into the acoustic streaming phenomenon in SAW devices. Fluid domain was modeled using the Navier Stokes' equation; the arbitrary Lagrangian Eulerian approach was employed to handle the mesh distortions arising from the motion of the solid substrate. The fluid was modeled as incompressible, viscous, and Newtonian. The fluid-solid coupling was established by maintaining stress and displacement continuity at the fluid-structure interface. The fluid mesh was continuously updated as the piezoelectric substrate undergoes deformation. A transient analysis was carried out by applying a time varying voltage to the transmitter IDT fingers. Results and discussion: We present the first report on a 3-D fluid-structure interaction (FSI) model of SAW sensor utilizing a multi-directional transducer based on a LGS substrate to investigate the streaming velocity fields and forces induced by SAW device. Figure 1 shows the schematic of the transducer configuration in the simulated SAW device, with IDT finger pairs defined on the surface for each port along Euler directions (0, 22, 0) and (0, 22, 90). The center frequency of the device calculated using an impulse response analysis was 68 MHz. Therefore, an AC analysis was carried out with a peak voltage of 2.5 V and frequency of 68 MHz applied to the transmitter IDT fingers. Our results indicate that pure shear horizontal (SH) mode propagates along the (0, 22, 90) direction making it suitable for biosensing (Fig. 2). Our work also suggests that the (0, 22, 90) direction possesses shallow penetration depth, thereby making it suited for liquid sensing applications, such as those in biosensing in bodily fluids. In addition, (0, 22, 0) direction shows the presence of mixed modes (Fig. 3). Despite the wave being mixed mode, surface normal component is the dominant component in the orthogonal (0, 22, 0) direction; this can be utilized for acoustic streaming. Further, a fluid-structure interaction model for the device with transducer along (0, 22, 0) direction was developed and solved

(Fig.4). The fluid velocity profiles obtained as a result of the fluid-piezoelectric device interaction indicate fluid recirculation in the regions close to fluid-device surface. Fluid motion, observed near the device surface, exerts stresses on the fluid-structure interfacial boundary which in turn induces the removal of loosely and non-specifically bound proteins from the device surface. Whereas the tangential components of the fluid velocity exert viscous shear stress creating a drag force, the normal component exerts a lift force. Our 3D FSI model results indicate that the tangential velocity component along the propagation direction is predominant suggesting that the drag force is significantly greater than the lift force and therefore, the particle advection induced by the former is the dominant mechanism of bio-fouling removal for the SAW sensor based on LGS. Our simulation results indicate that simultaneous sensing and non-specifically bound protein removal can be achieved through the use of multidirectional transducers on single piezoelectric device. Details will be presented at the AIChE 2009 meeting. References 1. J. J. Campbell and W. R. Jones, IEEE Transactions on Sonics and Ultrasonics 17, 71 (1970). 2. S. K. R. S. Sankaranarayanan, S. Cular, V. R. Bhethanabotla, et al., Physical Review E. 77, 066308 (2008)

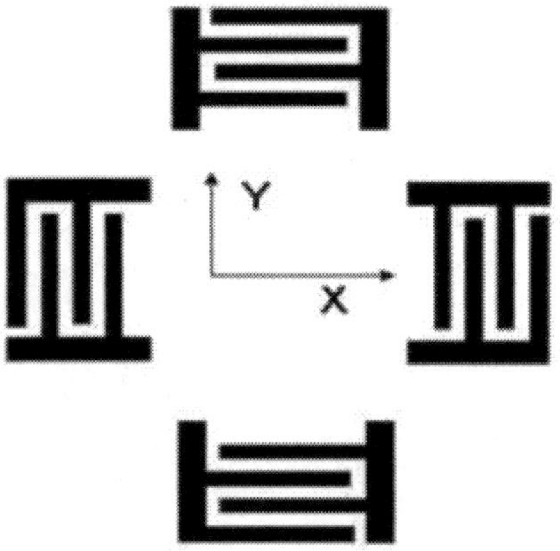

Fig. 1: Schematic diagram of an orthogonal surface acoustic wave device simulated in the current work. X and Y denote (0, 22, 90) and (0, 22, 0) directions in the Langasite substrate, respectively.

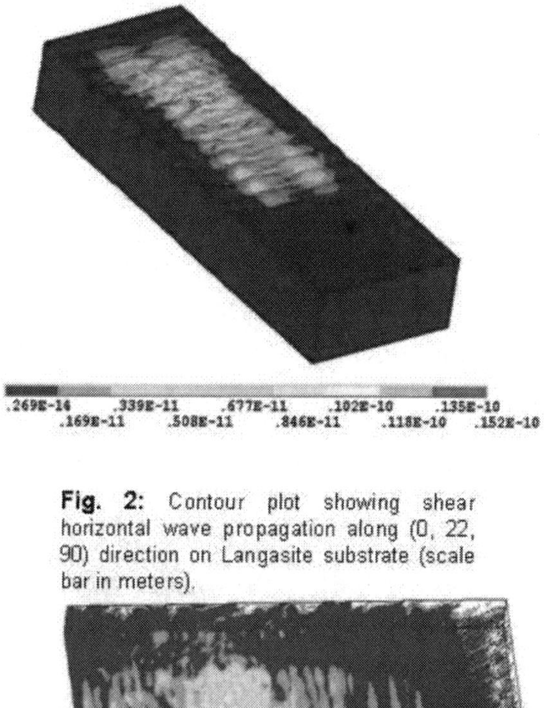

Fig. 2: Contour plot showing shear horizontal wave propagation along (0, 22, 90) direction on Langasite substrate (scale bar in meters).

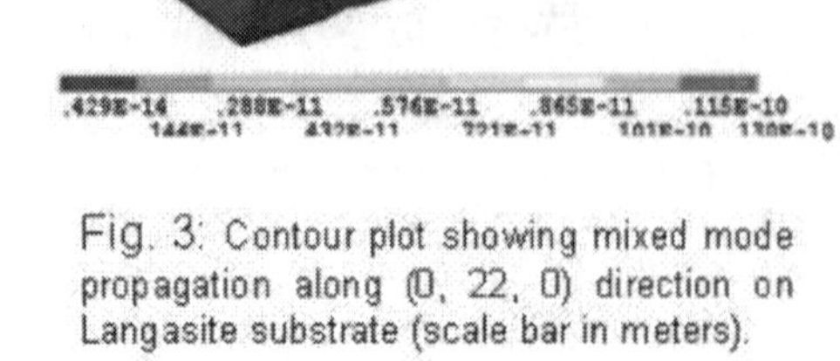

Fig. 3: Contour plot showing mixed mode propagation along (0, 22, 0) direction on Langasite substrate (scale bar in meters).

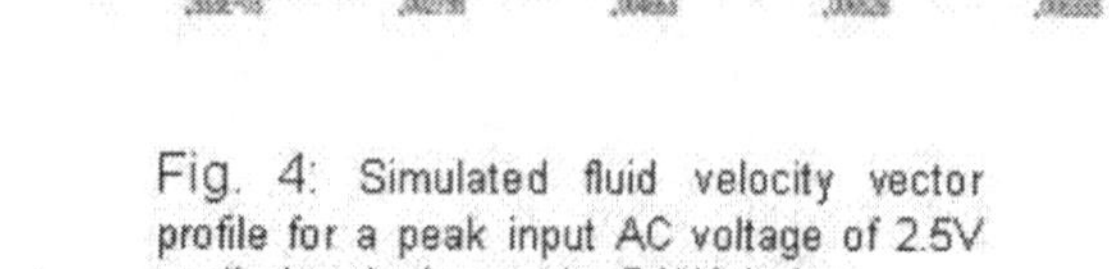

Fig. 4: Simulated fluid velocity vector profile for a peak input AC voltage of 2.5V applied to the Langasite SAW device

See more of this Session: Biosensor Devices II
See more of this Group/Topical: Topical 9: Sensors

Flow-Enhanced Performance of Optical Microcavity Biosensors

Jason M. Gamba, Chemical Engineering, California Institute of Technology, Pasadena, CA

Richard C. Flagan, Chemical Engineering, California Institute of Technology, Pasadena, CA

Whispering gallery mode optical resonators have become useful tools for label-free detection in recent years. While various geometries and methods of implementation have been studied, mass transport and flow around the optical microcavities has received relatively little attention. Here we report methods to optimize the interaction between the whispering gallery mode of the sensor and the species adsorbing to the surface by controlling flow around the resonator in order to improve the dynamic range. We model the flow conditions using COMSOL Multiphysics and compare these data with experimental results for a variety of resonator geometries.

See more of this Session: Biosensor Devices II
See more of this Group/Topical: Topical 9: Sensors

Photopatternable Hydrogel Material for Reversible Optical Hydrogen Peroxide and Glucose Sensors

Christian Pick, Department of Chemical and Biological Engineering, Missouri University of Science and Technology, Rolla, MO

Chang-soo Kim, Electrical & Computer Engineering, Missouri University of Science and Technology, Rolla, MO

David B. Henthorn, Department of Chemical & Biological Engineering, Missouri University of Science and Technology, Rolla, MO

In this work, a novel photopatternable hydrogel-based material for the fluorescent detection of hydrogen peroxide and glucose was developed and studied as a possible sensing element in optical glucose biosensors. This composite hydrogel material was developed to maximize analyte transport, be amenable to existing microfabrication techniques, and dovetail with various enzyme immobilization strategies. Nanoparticles loaded with a hydrogen peroxide sensitive europium tetracycline complex (EuTc) were mixed with monomer and crosslinker to form a photopolymerizable precursor. Sensitivity to glucose can be introduced through addition of methacrylated glucose oxidase into the precursor solution, allowing for covalent immobilization of the enzyme. As a result, the material can be integrated directly into optical biosensors for continuous glucose monitoring.

See more of this Session: Biosensor Devices II
See more of this Group/Topical: Topical 9: Sensors

Molecular Interactions in DNA and Morpholino Surface Hybridization

Ping Gong, Chemical Engineering, Columbia University, New York, NY

Kang Wang, Chemical & Biological Engineering, Polytechnic Institute of NYU, Brooklyn, NY

Napoleon Tercero, Chemical Engineering, Columbia University, New York, NY

Kenneth Shepard, Electrical Engineering, Columbia University, New York, NY

Rastislav Levicky, Chemical & Biological Engineering, Polytechnic Institute of NYU, Brooklyn, NY

Surface hybridization is widely encountered in diagnostic technologies for genotyping and gene expression analysis. In order to better understand the underlying fundamental processes, we compare two situations: (1) immobilized DNA oligonucleotides hybridizing to ones free in solution and (2) immobilized morpholino oligonucleotides hybridizing to free DNA oligonucleotides. Morpholinos, which obey standard base-pairing behavior, have a redesigned backbone chemistry that is charge-neutral. The influence of backbone charge is assessed from "two-color" hybridization electrochemical experiments in which surface populations of immobilized "probe" and hybridizing "target" species are simultaneously monitored. Hybridization thermodynamics in solution are analyzed for comparison. Qualitatively different behavior is encountered for the two scenarios of DNA or morpholino probes. For morpholino-DNA surface hybridization, probe coverage is a weak variable. As probe coverage increases, after an initial rise in target coverage due to elevation in surface binding capacity, hybridization becomes nearly independent of probe coverage. In contrast, DNA-DNA hybridization is highly sensitive to the immobilized probe, with strong suppression at high probe coverages. Both types of surface hybridization exhibit strong dependence on salt concentration; in the morpholino-DNA case this sensitivity is attributed to accumulation of repulsive interactions between hybridized target strands. Interestingly, morpholino-DNA hybridization in solution is independent of salt concentration in the range from deionized water to 1 M buffer. Comparison of experimental data with physical models incorporating electrostatic and associative strand interactions suggests influence of attractive strand-strand and strand-surface interactions.

See more of this Session: Biosensor Devices II
See more of this Group/Topical: Topical 9: Sensors

The Rational Design of Nitric Oxide Selectivity in Single-Walled Carbon Nanotube near Infrared Fluorescence Sensors for Biological Detection

Jingqing Zhang, Chemical Engineering, MIT, Cambridge, MA

Jong-Ho Kim, Chemical Engineering, MIT, Cambridge

Hong Jin, MIT

Paul W. Barone, Chemical and Biomolecular Engineering, MIT, Cambridge, MA

Daniel A. Heller, Chemical Engineering, Massachusetts Institute of Technology, Cambridge, MA

Changsik Song, Chemical Engineering, Massachusetts Institute of Technology, Cambridge, MA

Michael Strano, Chemical Engineering, Massachusetts Institute of Technology, Cambridge, MA

A major challenge in the synthesis of nanotube or nanowire sensors is imparting selective analyte binding through means other than covalent linkages which compromise electronic and optical properties. We synthesize a 3,4-diaminophenyl-functionalized dextran (DAP-dex) wrapping for single-walled carbon nanotubes (SWNT) that imparts rapid and selective fluorescence detection of NO, a messenger for biological signaling. The near infrared (nIR) fluorescence of SWNTDAP-dex is immediately and directly bleached by NO, but not by other reactive nitrogen and oxygen species. This bleaching is reversible and shown to be caused by electron transfer from the top of the valence band of SWNT to the lowest unoccupied molecular orbital (LUMO) of NO. The resulting optical sensor is capable of real-time and spatially resolved detection of NO produced by stimulating NO synthase (iNOS) in macrophage cells. We also demonstrate the first optical sensor capable of in-vivo NO detection in a mouse model.

See more of this Session: Biosensor Devices III
See more of this Group/Topical: Topical 9: Sensors

Fiber Optic Enzymatic Biosensor Array for Measurement of Halogenated Hydrocarbon Mixtures Using a Chemometric Approach

Zhong Zhong, Department of Chemical and Biological Engineering, Colorado State University, Fort Collins, CO

Weina Wang, Electrical and Computer Engineering, Colorado State University, Fort Collins, CO

Kevin L. Lear, Electrical and Computer Engineering, Colorado State University, Fort Collins, CO

David S. Dandy, Department of Chemical and Biological Engineering, Colorado State University, Fort Collins, CO

Bernd Hitzmann, Institute of Technical Chemistry, Leibniz University of Hannover, Hannover, Germany

Kenneth F. Reardon, Department of Chemical and Biological Engineering, Colorado State University, Fort Collins, CO

Halogenated hydrocarbons are among the most common environmental contaminants in groundwater, due to their widespread use and their recalcitrance to natural attenuation. Halogenated alkanes (1,2-dichloroethane, 1,2-dibromoethane) could cause health problems and could be found in most environmental groundwater monitoring sites. Traditional laboratory-based analytical methods for these halogenated hydrocarbon mixtures is usually time-consuming and expensive. Biosensors could avoid these disadvantages by providing real-time in-situ measurement with highly specific and sensitive detection. We have developed a biosensor platform that includes a biological component that detects the analyte through an enzymatic reaction, a transducer to translate the biological signal into an optical signal, and a detector that recognizes the signal and correlates it to the concentration of analyte. Recently, we have developed a biosensor array consisting of multiple biosensors with different enzymes that have different activity toward the analytes in a mixture. Chemoemtric approaches are used to process these data.

In this study, a pH-based biosensor array system has been used to demonstrate the biosensor array and chemometric approach for analysis. The individual pH-based biosensors are made by using dehalogenase-expressing bacteria as the biocomponent and pH optodes as transducer. The reaction of the dehalogenases with the analytes generates protons and the pH optodes detect the resulting pH change. A set of dehalogenases is applied in different biosensors in order to build up a pH-based biosensor array, while sequential calibration on individual biosensor may be required prior to measurement. The biosensor array was capable of resolving mixtures of halogenated hydrocarbons with good accuracy. Results with different calibration methods and chemometric approaches will be discussed.

See more of this Session: Biosensor Devices III
See more of this Group/Topical: Topical 9: Sensors

Multiple Biomarker Detection Using SH-SAW Resonator Array

Mandek B. Richardson, Chemical & Biomedical Engineering, University of South Florida, Tampa, FL

Venkat Bhethanabotla, Chemical and Biomedical Engineering, University of South Florida, Tampa, FL

Darren W. Branch, Biosensors and Nanomaterials Department, Sandia National Laboratories, Albuquerque, NM

Thayne L. Edwards, Biosensors and Nanomaterials Department, Sandia National Laboratories, Albuquerque, NM

Summary

The specific detection of multiple biomarkers to assess disease state is critical before starting any treatment protocol. Though surface acoustic wave sensors have been successfully used to detect biomarkers and antigens, implementation has been limited to serial detection of typically one biomarker at a time. For simultaneous biomarker detection, we have developed a shear-horizontal surface acoustic wave (SH-SAW) array that uses six resonators each operating at 408 MHz. Each channel uses two-port resonators operating on 36° YX lithium tantalate (36° Y LTO) for use in fluidic environments. Crucial to this work was the development of a fluidic interface that permits surface bio-activation yet is easy to assemble while maintaining sufficient quality factor (Q) of the entire sensor assembly. The second major contribution is the development of a custom oscillator circuit that monitors the frequency shift of each channel simultaneously while minimizing channel cross-talk.

Motivation

To date nearly all SHSAW biosensors are configured as delay-lines for biological detection applications [1,2]. A major obstacle for acoustic biosensors is reducing attenuation due to viscous and capacitive loading as this impacts overall sensitivity. A major difference between delay-line devices and resonators is that even with high insertion loss the phase response is often preserved in delay-line devices whereas resonators cease to operate. For this reason SHSAW resonators must maintain a high Q with low acoustic loss packaging. In contrast to our previous work which used a four channel SHSAW array delay-line device [3], resonators offer the advantage of higher sensitivity, smaller package with multiple channels, oscillator based measurements, and high reproducibility. For use as commercial sensors, resonators offer greater advantages due to their superior reproducibility.

Early detection of ovarian cancer is imperative to ensure the patient's likelihood of survival. At this time a lack of early symptoms and the absence of a reliable screening test results in 70% of women being diagnosed after the cancer has spread beyond the ovary. Currently there are tests that screen for a single biomarker, CA-125. However, the current methods have proven to be unreliable and inaccurate. For example, increased serum levels of CA-125 are found in only half of the patients diagnosed with stage I ovarian cancer. The discovery of three novel biomarkers combined with the ability to simultaneously test them with the SH-SAW resonator array device, will be invaluable in establishing a definitive, reliable, accurate and economical test for early stage ovarian cancer.

References

1. Tamarin, O., S. "Real time device for biosensing: Design of a bacteriophage model using love acoustic waves." Biosensors and Bioelectronics 18: 755-763, 2003.

2. A. Malavé, U. Schlecht, T. M. A.Gronewold, M. Perpeet, and M. Tewes, "Lithium tantalate surface acoustic wave sensors for bio-analytical applications," IEEE Sensors, pp. 604-607, 2006.

3. D. W. Branch and T. L. Edwards, "Love Wave Acoustic Array Biosensor Platform for Autonomous Detection," in IEEE Intern. Ultrasonics Sym. New York, NY, 2008, pp. 260-263.

See more of this Session: Biosensor Devices III
See more of this Group/Topical: Topical 9: Sensors

Liposome Microarrays for Toxin Detection

Isabel Sole Font, Department of Chemical Engineering, City College, New York, NY

Vilobh Sete, Department of Chemical Engineering, City College, New York, NY

Jude Phillip, Department of Chemical Engineering, City College, New York, NY

Alexander Couzis, Department of Chemical Engineering, City College, New York, NY

The technology of protein microarrays (or "protein chips") is a recent innovation used for screening of the binding interactions of proteins (e.g. an antigen, a cell membrane receptor, an enzyme) with potential binding partners or ligands. The ligands can be other proteins (e.g. antibodies or hormones), enzyme substrates, or small synthetic molecules such as oligopeptides, or their analogues (peptoids, etc.) prepared as part of a combinatorial chemistry library. In this work, we have developed a new microarray platform for arraying membrane receptors for multiplexed screening of the binding interactions of these receptors with binding ligands. We used interferometric lithography to make large areas of periodic photoresist structures on a Si wafer for use as templates for metal deposition. Following metal deposition, the photoresist was removed, leaving ordered metal arrays on a flat Si surface. The bottom of the wells was subsequently functionalized with an amine termination by backfilling with an amine silane. The amine functionalized wells were then conjugated to Neutravidin labeled with a FITC dye. As a proof of concept experiments, we used the intact liposome microarray format and known receptor-ligand system to check the applicability of the liposome microarray platform for biodetection. Monoganglioside GM1 was incorporated in the lipid bilayer of liposomes and these receptor liposomes were arrayed onto the chemically functionalized microwell patterned substrates. These GM1 receptor incorporated liposome arrays were used to detect specific binding of Cholera Toxin Subunit B (CTB) to the surface receptors using confocal laser scanning microscopy.

See more of this Session: Biosensor Devices III
See more of this Group/Topical: Topical 9: Sensors

Identification of Troponin I Binding Peptides Using Phage Display for Biosensor Development

Jong Pil Park, Chemical Engineering, Columbia University, New York, NY

Donald M. Cropek, Construction Engineering Research Laboratory, US Army Engineer Research and Development Center, Champaign, IL

Scott Banta, Chemical Engineering, Columbia University, New York, NY

Acute myocardial infarction (AMI) is the leading cause of morbidity and mortality in most industrialized nations throughout the world. Several clinical biomarkers are commonly used for the early detection of heart disease including the MB isoenzyme of creatine kinase (CK-MB), and myoglobin, however, there are some limitations to their use due to low specificity, selectivity and sensitivity. Troponin I has rapidly become the preferred diagnostic biomarker due to its high specificity and selectivity for myocardial injury.

We have used polyvalent phage display to isolate unique linear peptide motifs that recognize both the human and rat homologues of troponin I. The peptide specific for human troponin I has a sequence of FYSHSFHENWPS and the peptide specific for the rat troponin I has a sequence of FHSSWPVNGSTI. Enzyme-linked immunosorbent assays (ELISAs) were used to evaluate the binding interactions, and the two peptides exhibited some cross-reactivity, but they were both more specific for the troponin I homologue they were selected against. The binding affinities of the peptides were impacted in the presence of complex tissue culture media (MEM). And the addition of 10% calf serum further interfered with the binding of the target proteins. Kinetic indirect ELISAs revealed that both troponin I binding peptides were found to be nanomolar binders while attached to the phage particles. Specifically, the K_d value of peptide selected against rat troponin I was 0.95 ± 0.09 nM. The peptide obtained through selection against human troponin I had a K_d value of 2.5 ± 0.12 nM. These new peptides will potentially have utility in the development of new clinical assays for cardiac injury as well as in monitoring of cardiac cells in tissue culture applications.

See more of this Session: Biosensor Devices III
See more of this Group/Topical: Topical 9: Sensors

Novel Bacterial Biosensors for the Discovery of Subtype-Selective Thyroid Receptor Modulating Drugs

Jingjing Li, Department of Chemical Engineering, Princeton University, Princeton, NJ

Izabela Hartman, Department of Chemical Engineering, Princeton University, Princeton, NJ

Thomas William Eyster, Department of Chemical Engineering, Princeton University, Princeton, NJ

David W. Wood, Department of Chemical Engineering, Princeton University, Princeton, NJ

Thyroid hormone receptors (TRs) belong to the nuclear hormone receptor superfamily, and play a critical role in heart metabolism and adipogenesis through thyroid hormone signaling. Two subtypes of TR, α and β, exhibit obvious organ preference and are thought to play different roles in thyroid-based metabolic regulation. The discovery of subtype-selective therapeutics for the TRs is therefore a critical goal for this important drug target. This, as well as high-throughput detection of TR-targeting endocrine disruptors, calls for a subtype-distinguishable biosensor platform. Our lab has developed a modular generic biosensor where a desired nuclear hormone receptor (NHR) gene is fused to an engineered intein and thymidylate synthase (TS) reporter enzyme. Expression of this fusion in TS deficient E. coli leads to an NHR ligand-dependent growth phenotype. In the present work, two new bacterial biosensors for thyroid hormones and analogues were constructed by insertion of either the TR α 1 or TR β1 genes. These new biosensors show phenotype dependence to the natural TR ligand, 3,3',5'-triiodo-L-thyronine (T3). Further, a variety of synthetic ligands were found to cause a differential responses in the TR α1 and TR β1 sensors, suggesting these sensors are capable of distinguishing subtype-selective ligands. In addition, the sensors were able to distinguish agonist from antagonist compounds via the promotion or inhibition of cell growth. Finally, the half-maximal effect concentrations (EC50) for some ligands were measured and found to be consistent with their reported affinities to the TRs. Thus our results show that these TR α1 and TR β1 biosensors can act as a powerful tool for the discovery of subtype-selective drugs for TR-related disorders.

See more of this Session: Biosensor Devices III
See more of this Group/Topical: Topical 9: Sensors

AUTHOR INDEX

AUTHOR INDEX